T0094351

SUSHI

Food for the
eye, the body
& the soul

Ole G. Mouritsen

SUSHI
Food for the eye, the body & the soul

Graphic design and photography
Jonas Drotner Mouritsen

Water colours
Tove Nyberg

Translation and adaptation to English
Mariela Johansen

🜨 Springer

Sushi • Food for the eye, the body & the soul

Author
Ole G. Mouritsen

Graphic design and photography
Jonas Drotner Mouritsen

Water colours
Tove Nyberg

Translation and adaptation to English
Mariela Johansen

ISBN: 978-1-4419-0617-5   e-ISBN: 978-1-4419-0618-2

Library of Congress Control Number: 2009931802

Printed in the US on acid-free paper

9 8 7 6 5 4 3 2 1

springer.com

Menu

• • •

To
Myer Bloom
for first introducing me to the world of sushi

&

to
sushi chefs around the world
whose art, craft, and aesthetic sense
keep me spellbound

Contents

❧ A detailed explanation • ❧ A narrative detour

Tools, preparation & presentation

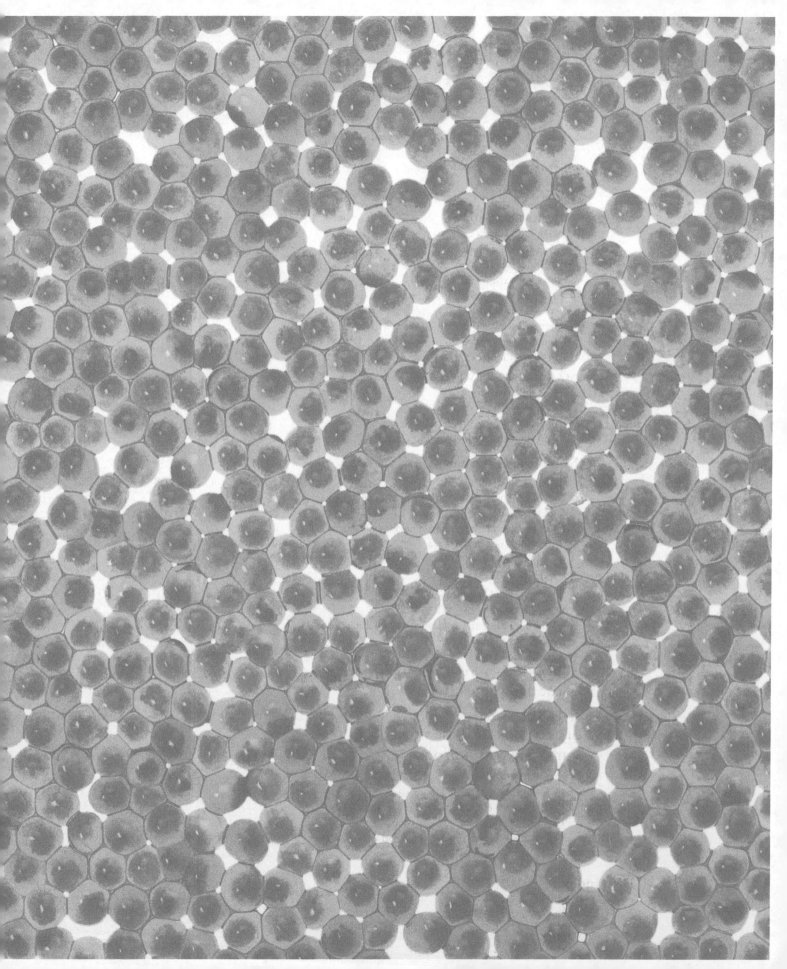

Preface

This book unfolds the story of my ongoing love affair with sushi. I began to write it at Christmas 1996, a time when I had already acquired a fair bit of experience with eating at sushi bars and preparing dishes at home. It was a source of puzzlement to me that there were so few sushi restaurants in the part of the world where I live and that Scandinavians were so reticent about Japanese food in general, and raw fish in particular. This latter point is notable, because traditional Nordic cuisine incorporates sushi-like uncooked elements such as marinated herring and cured salmon ('gravlaks'). It also seemed remarkable that sushi had not yet become fashionable as it is common knowledge that food from the sea is nourishing and promotes wellness. It is generally acknowledged that the Japanese diet may be the key to a long and healthy life.

I made it my goal to write a sort of primitive 'cookbook' in order to convey to my fellow citizens my passion for sushi and the Japanese way of preparing food. But this project languished untouched in my filing cabinet drawer for a long time. In the meanwhile, the number of sushi restaurants and take-outs virtually exploded in the West, even in Scandinavia. It has become trendy, particularly among urban professionals and young people, to eat sushi. For the vast majority of the population, however, sushi is still something that is prepared by others or that one picks up at the supermarket deli counter. Furthermore, some sushi restaurants have put their offerings on a pedestal, implying that it is only for the *cognoscenti*, far from its humble origins as an everyday food. One can still encounter overbearing servers who try to intimidate the uninitiated – trendy dining can have its drawbacks.

This is truly regrettable. Although still considered exotic by many, sushi is a very ordinary food with endless possibilities for variations that are not necessarily tied to rigid traditions or special and esoteric ingredients. It was this conviction that led me at last to retrieve the book project from the filing cabinet in an attempt to shed a different light on my favourite food.

Some sushi raw ingredients: avocados, sushi rice, seaweed, raw salmon, and a piece of cooked octopus.

I also have a secondary motive for writing this volume, namely, my interest in curiosity driven questions concerning raw foodstuffs, their preparation, and the tools used to do so. What makes fish, especially raw fish, such a healthy food? What happens when one cooks or ferments rice and soybeans? Why do crustaceans turn red when cooked? Which chemical compounds in the *shiso* plant enable it to act as a preservative for plums? How is a superior fish knife sharpened? Why are wooden tools better than plastic ones? These questions, and many others, refer to the fundamental science behind sushi, *sashimi*, and other Japanese food, which we could call sushi science. As a scientist I cannot help being amazed at the small miracles which take place in the kitchen.

How and why do they occur, and what do they mean? The joy of discovering answers to these puzzles simply transforms the process both of preparing and of consuming a meal – it adds an extra dimension of pleasure.

There is an expanding professional interest among chefs and nutritionists in the science underlying the preparation of food. The term 'molecular gastronomy' has recently been coined to describe the study of the molecular properties of food ingredients and their transformation in the course of their journey to the table, as well as associated attempts to provide quantitative explanations for flavour and taste sensations. Because the discipline is still in its infancy, current molecular gastronomy is often not based on a sufficiently scientific understanding at the molecular level while, on the other hand, some chefs complain that it places too little emphasis on true gastronomy. As an extension of the scientific aspect of this field, my colleagues and I have in the last few years attempted to develop an insight into the physics and the physical chemistry inherent in food and its preparation, a field which we have dubbed gastrophysics.

Hence, this book is not really a 'cookbook' in the ordinary sense; rather, it takes a holistic approach to eating a sushi meal and the enjoyment inherent in preparing it – a meal which not only nour-

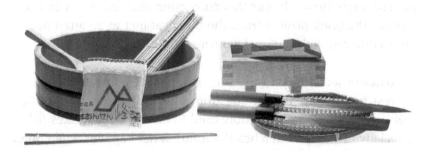

Some of the tools used in the preparation of sushi.

ishes the body but also enriches the brain and delights the senses. It revolves around a type of food in which the quality of the raw ingredients, their taste, the chemical composition, the physical texture, and the overall aesthetic impact are inseparable and equally important entities. This book tries to write its way to a deeper understanding of what gastrophysics could become.

Sushi is fish or shellfish on balls of vinegared, cooked rice.

A number of people have been important for me in my journey towards the completion of this book. First, I am grateful to my mentor, collaborator, and friend of many years' standing, Professor Myer Bloom, who introduced me to sushi in Vancouver in 1980, both at the sushi bar and in his home. My wife Kirsten and our children, Julie and Jonas, through their never wavering interest in, and insatiable appetite for, sushi have been a driving force behind my experimentation with the preparation of the family's weekly sushi dinner. Throughout these years, sympathetic fish sellers and sushi chefs have given me consistent and indispensable support. Most recently, contact with Claus Skovsted and Søren Gordon at *bar'sushi* and *Goma* in Odense, Denmark, have been a source of inspiration. Over and above this, Claus and Søren generously prepared sushi for some of the illustrations in the book. Head chef Jacob Jo Jørgensen cooperated in a photo session at *Sticks 'n' Sushi* in Copenhagen. Yoshikata Koga and Motomu Tanaka kindly assisted me with the use of *kanji*, the Chinese characters used in the modern Japanese logographic system. Carl Th. Pedersen undertook critical proofreading with particular reference to chemistry-related aspects of the book. Midori Fischer from *Nihonjinkai*, The Japanese Society of Denmark, gave me invaluable guidance and help in the correct use of Japanese expressions. Julie Drotner Mouritsen, Kirsten Drotner, Per Lyngs Hansen, Ulla Lauritsen, and Amy Rowat read through various versions of my original manuscript and provided me with good and constructive feedback. I owe thanks to the members of The Gastrophysical Society who have shared with me their vast interest in delving into the scientific mysteries of food. I am indebted to my Springer editor Maria Bellantone for her instant and continuous enthusiasm for the project of producing an English edition of the book. Finally, my sincere thanks go to my good friend Mariela Johansen, who shares my passion for sushi and who took the initiative to translate the book into English. She has undertaken this task with an admirable scrutiny and professionalism and the book has greatly improved in her hands.

学術 THE BLUE *KANJI* CHARACTERS identify the sidebars in which you can look for detailed, and often more specialized, information that reinforces the main text. In some cases, this is of a more technical or scientific nature, but it is not a precondition for an understanding of the primary narrative. The symbols mean science and scholarly endeavours.

珍談 THE GREEN *KANJI* CHARACTERS identify the sidebars in which you can read anecdotes and folkloric explanations that are entertaining, but possibly spurious. The symbols stand for an amusing and odd story that should be taken with a grain of salt.

調理法 THE RED *KANJI* CHARACTERS identify the sidebars in which you will find instructions and recipes for the preparation of sushi, complementary side dishes, and condiments. The symbols are the words for recipe and a description of how to prepare food.

In a totally different way, I am deeply and humbly grateful to the countless chefs working in sushi bars all around the world who indulgently let me look over the counter to 'steal from the master', allowing me to tease out their secrets and providing me with helpful explanations. It is no wonder that my wife always says that I honed my sushi skills the expensive way!

I would also like to acknowledge, with gratitude, THE VILLUM KANN RASMUSSEN FOUNDATION which made it possible for me to undertake a two-week working stay at San Cataldo in Italy in May 2005 by bestowing on me its annual award for technical research. It was in the library of this former convent, overlooking the beautiful blue Amalfi Bay, that many parts of this manuscript saw the light of day.

ABOUT THE BOOK AND HOW TO READ IT

The book consists of four story lines interwoven with each other to form a many-faceted whole, but written in such a way that they can still profitably be read as independent entities.

The first story line treats the raw ingredients used in the preparation of sushi and the various side dishes which complement it. Here you will learn about the plants, algae, and animals from which the ingredients are derived, with particular emphasis on their texture, taste, and nutritional content, e.g., proteins, carbohydrates, and fats. You can also discover something about the scientific underpinnings of these component materials and how their proper treatment can enhance flavour, texture, and appearance.

The second story line deals with the various tools that are used in the preparation of sushi.

The third story line describes the preparation of sushi and its complementary dishes and accompaniments, using my own experiences as the point of departure. This part of the book is intended more as a catalogue of inspirational ideas than as a systematic exposition on sushi preparation. Here you will find some simple instructions which, with a little practice, any interested person could carry out at home. The approach is *Zen*-inspired: first learn a few rules and basic facts, practice them to gain proficiency, and then forget everything about the rules and let your intuition and imagination be your guide. The instructions are accompanied by illustrations that serve to accentuate the aesthetic dimensions of the preparation, creation, and presentation of a sushi meal.

The fourth story line consists of a number of small stories, essays, and anecdotes linking sushi and Japanese cuisine with aspects of cultural history, wellness, and science.

Many of the factual explanations about the science associated with the raw ingredients build on McGee's wonderful book *On Food and Cooking: The Science and the Lore of the Kitchen* (2004) and Belitz et al.'s encyclopedic *Food Chemistry* (2004). Both books are recommended to those readers who wish to delve deeper into the subject.

As a general rule, the book deals only with sushi which can be prepared from ingredients that are generally available outside of Japan, although finding them may occasionally entail a trip to a specialty shop. The exceptions to this rule occur in those cases where they reinforce a general principle or simply permit the telling of a good story.

The majority of the Japanese words are indicated by *italics* and they are defined in a glossary at the back of the book. One exception is 'sushi' (or 'zushi'), which has found its way into many languages. Similarly, Latin names and scientific and technical terms are italicized when they first appear in a given context. These terms are likewise explained in a list of scientific terminology in the same section.

At the back of the book you will also find a bibliography which lists the written sources that provided background material for the writing of the book. It includes titles dealing with sushi and Japanese food, the science of cooking, as well as cultural history, nutrition, and wellness. I generally have omitted references to professional journals.

学術 SUSHI OR ZUSHI? Sushi and zushi have the same meaning, but are pronounced with an s or a z sound, respectively. Strictly speaking, in Japanese sushi is pronounced with a voiced z sound when it is linked to another word. For this reason, 'nigiri-sushi' is pronounced and written as 'nigiri-zushi', but zushi is not a word in its own right. A similar rule applies to other Japanese words which start with an s.

"Irrasshai!"

"Irasshai, irasshaimase!" This is the sound of the sushi chef's salutation and an invitation to come on in when you arrive at the sushi bar. Either at the entrance or on the counter of most sushi bars, another greeter awaits the guests, *Manekineko*, the Japanese good luck cat. This little smiling figure, which is holding one paw in the air, its way of saying welcome, conveys a double meaning. If the left paw is raised, it is welcoming the guests. If the right paw is raised, it is welcoming money and prosperity to the business establishment. *Manekineko* attracts wealth and success to the owner.

Once you are well inside the door, a decision awaits you – whether to sit at the bar or use a table or booth. A group often prefers to be seated at a table or perhaps, if the restaurant has one, a separate *tatami*-room with woven bamboo mats. The true sushi gourmet will instantly opt for a spot at the bar, preferably a place where he or she can closely follow the artistry of the sushi chefs as they work.

If it is the first time you are in a sushi restaurant, this will probably be obvious to the server and the chef and they might discreetly try to ask you what you would like. My advice, if you really want to try sushi, is to choose a place at the bar – in my opinion, absolutely the best way to experience the very special atmosphere associated with a sushi meal. This is especially the case if you are in Japan and do not understand the language and would have difficulty ordering or reading a menu – if there is one.

Sometimes the different dishes and the specials of the day are posted on a small wooden board above the bar. Sitting at the counter you will have the advantage of being able to point to what you would like, which is probably to be found in the refrigerated glass case built into the bar. Should you not wish to order à la carte and prefer an existing combination order, you can still position yourself at the bar and observe the intricate interplay between the chef and his assistants. If you feel adventurous and are prepared to issue a challenge to the sushi chef, you can ask for *omakase*. Then he is at complete liberty to decide on the composition of the sushi meal – but be warned that you may lose control of the bill and will have to pay, no matter how much it ends up costing.

After you are seated, the waiter will come with a little wooden tray with a rolled up white washcloth – steaming and warm – so that you can refresh your face and hands. It is the accepted custom to use the cloth in moderation, just as it is considered polite to roll the cloth up again before putting it back on the tray. At this point, you place your order either with the server, directly with the chef, or, in some places, fill in a form which lists the available sushi. You can always ask if there is something which is particularly recommended on a given day. Also, make sure you inform yourself about the prices if they are not listed on the menu. Orders for hot dishes, such as *miso* soup or *tempura*, which are prepared in the kitchen, are normally given to the server. He or she will also bring the drinks, typically green tea, beer, and *sake*. At that point you can settle in to enjoy the dual pleasures of watching the chefs perform and of savouring the food.

It is no exaggeration to say that this is performance art on a grand scale. Skilled sushi chefs are culinary artists who are engaged in a balancing act with their knives, chopsticks, and bamboo mats. While it looks like child's play, achieving this level of perfection requires a long apprenticeship and much training – at least five years. According to an old Japanese saying, it takes seventeen years to become a true sushi master.

In addition, it is easy to overlook all the preparation that takes place in advance or behind the scenes by assistants in the kitchen. If you want to gain an insight into the amount of work involved, it is a good idea to come to the sushi bar early in the day, if they are open for customers, avoiding the busy lunch and dinner hours. At your leisure you will be able to see the sushi chef cut up whole fish, crack open crabs and clams, and reduce radishes to delicate mounds of ultra-thin ribbons. If you are lucky, you may even find out how one sharpens a genuine sushi knife.

One is never bored in a sushi bar. The occasional wait during the busy periods fly by if one tunes in to the rhythm with which the chefs pass one fantastic dish after another over the counter. Taking notice of the choices of the more experienced sushi eaters can be a source of inspiration and the chefs are normally happy to replicate them for you. Even if you have ordered a whole meal in one go, the dishes arrive at intervals so that they are always fresh. The chef is often looking after several customers at once, and it is amazing to observe at first hand how adept a good sushi chef is at remembering exactly who ordered what.

All sushi bars are, in principle, organized in the same way. The chef normally stands in a small space in the room with his back to a wall that has shelves with the special trays on which the dishes are presented. One end of the bar has access to the kitchen where the hot dishes are prepared.

The only heat source found in the bar is a gas grill or a small oven, which is used to warm such items as eel or to toast salmon skin. The bar is like a high counter with a wide bench on the inside with sinks and running cold water. Placed on it lengthwise is a long glass-enclosed refrigerated case in which fish, shellfish, and vegetables are carefully laid out on small trays, ready to be fetched in accordance with what the customer has ordered.

On the customers' side of the bar there is a raised countertop and the seating often consists of barstools. In a classical sushi bar, the countertop is made of cedar wood and the meal is served directly on it. Sometimes the surface is slanted to act as a sort of plate. When the chef is about to serve a dish, he first wipes the countertop with a clean wet rag and then places the newly created pieces of sushi directly on it, after which the customer quickly eats them.

More sophisticated sushi bars are arranged in such a way that the customers sit on several sides and the chefs work in the middle. A particularly high tech design is known as *kaiten*-zushi, where a small conveyor belt circulates past diners seated at a three- or four-sided counter. The chefs place a variety of sushi dishes on the constantly circulating belt from which the customers help themselves – a faster and less expensive way to eat than ordering à la carte. Plates are colour coded and the bill is tallied up by the server based on the total consumed. An amusing variation on this is 'love-boat sushi', which utilizes small wooden boats, each laden with a cargo of sushi, that sail around in a canal built into the bar.

SUSHI – *ZEN*, PASSION, SCIENCE & WELLNESS

On this plate of food
I see the whole universe behind my existence
A *Zen* mealtime saying

SUSHI AND *ZEN*

Sushi is a food that nourishes the body, enriches the brain, and is a delight for the eye. Sushi is a healthy food, in which the quality of the raw ingredients, the taste, the chemical composition, the physical texture, and the aesthetic presentation are inseparable entities. Sushi is a food where the pleasure taken in its preparation and the artistry of the presentation are just as important to the whole experience as the meal itself. Sushi encompasses passion, science, and wellness. Sushi is *Zen*.

A CONFESSION

Before going any further, I may as well come clean: sushi has literally become a consuming passion. When I visit a new town and am desperately looking for a sushi bar, I can suddenly relate to why an alcoholic craves a drink or a chain smoker has to have a cigarette. While life may sometimes seem like a series of minor accidents and difficulties, these moments of unadulterated indulgence that we allow ourselves are the spice of life.

The lengths I will go to in search of sushi can be illustrated by an old *Zen* legend which recounts the story of a wandering monk who encounters a hungry tiger. The tiger chases the monk onto the ledge of a cliff. To escape the tiger, the monk leaps over the side, but he manages to grab the branch of a tree which is growing on the slope just under the ledge. While the monk is dangling precariously, he catches sight of another hungry tiger standing below the cliff, patiently waiting for him to fall. As the monk's strength is failing him, he spies a wild strawberry growing just within his grasp. He lets go of the branch, picks the strawberry, and puts it in his mouth, fully aware that this is the last morsel he will ever eat. How sweet was the taste of that strawberry in that fleeting moment!

THE *HAIKU* MOMENT

The Japanese characterize this as a *haiku* moment – a special, brief second when one has a sensation of great insight and enlightenment and at the same time is aware of the transitory nature of the material world. The moment expresses love of life, while accepting that it inevitably must come to an end.

Haiku is a Japanese form of verse which can be described as being both the most minimalist and that which is most constrained. The following is a well known *haiku*:

> The old pond
> a frog jumps in
> sound of water

The verse was written by Matsuo Bashō (1644-1694), considered by many to be the most famous Japanese *haiku* poet. The line in this poem that is the key to its meaning is the last one – just three small words.

学
術 *HAIKU* IS A VERSE FORM that arose in Japan at the end of the 15th Century and which rests on a set of strict rules that have been expanded and refined over time.

Among the classical rules for the writing of *haiku* is the requirement that a poem shall consist of three lines with five, seven, and five *onji* (Japanes symbol-sounds, analogous to syllables), respectively. The poet can employ techniques such as similes, contrasts, riddles, narrowing of focus, ambiguity, word play, *sabi*, *wabi*, paradox, and humor.

The essence of *haiku* is bound up with *Zen* philosophy, even though poetry expressed in words in a sense stands in contradiction to the *Zen* concept that words act to inhibit a true understanding. In *Zen*, there are many little flashes of enlightenment, *kensho*, when a deeper reality can be glimpsed in the ordinary things of the world. In a sense, the *haiku* moment is the literary embodiment of *kensho*.

The best known early *haiku* poets are Matsuo Bashō (1644-1694), Yosa Buson (1716-1783), and Kabayashi Issa (1763-1827). A later poet is Kijo Murakami (1865-1938). Examples of *haiku* written by these authors are found in several places in this volume.

"Something from the sea and something from the mountains". Woodblock print by Utagawa (Ando) Hiroshige (1797-1858).

Jane Reichhold, an American authority on *haiku*, compares *haiku* to fish. According to her, the poems do not come from the writer, but come through him or her. In a similar manner, the *haiku* moment one can experience in turning a fish into sushi and eating it does not originate with the chef; he is merely the agent. The sublime aspect of the experience is particularly found in the *ku* (the short fragment of the poem) or in the appearance and taste of the fish, which cannot be described or explained in words.

Haiku have been characterized as poems composed of that which is only half said. But as Matsuo Bashō says: is there any need to say it all? The same holds true for this book on sushi. In it you will be able to read something about sushi, but not everything, and you will have to seek more knowledge for yourself and acquire your own expertise.

The composition of *haiku* is governed by many complicated rules concerning content and syntax and, over time, there have been various schools which have argued vigorously in defence of their perception of the correct way to write these verses. Matsuo Bashō gives the following advice: learn the rules and then forget all about them. This builds on the *Zen* way of thinking that enlightenment and insight do not come from the acquisition of knowledge, but rather from 'unlearning' it.

This is the concept on which I have built this book. There is an abundance of widely varying prescriptions for the right way to prepare sushi and different chefs and traditions all champion their particular points of view. In addition, the internationalization of sushi culture has had a paradoxical effect. On the one hand, it has helped to introduce new ingredients and techniques but, on the other, it has imposed a rather rigid conception of what constitutes sushi. As the latter is contrary to the original idea of sushi, my advice to the reader is as follows: read this book or others on the subject of sushi, learn a few rules and recipes, and then forget them all and let your intuition take over. Seize the *haiku* moment!

THE SCIENCE BEHIND THE PASSION

There is a common saying in Japan that every meal should incorporate "something from the sea and something from the mountains". The ocean supplies fish, shellfish, and seaweed, while rice, beans, and other plants come from the mountains.

My passion for sushi is grounded in a fascination with how one can use simple, healthy, and pure raw ingredients to compose a meal that will inspire a *haiku* moment. The intensity of this fleeting instant is enhanced by the rediscovery of the beauty inherent in the humblest things.

It is here that we encounter the science behind the passion. Here too, I must unburden myself by making another confession. I cannot help being amazed time and again by the small miracles that occur in a sushi bar and in a kitchen when one prepares food. Fish, seaweed, and vegetables are wonderful to behold and it is awe-inspiring to witness their transformation in the hands of a skillful chef. Why do the raw ingredients look the way they do and what actually happens when we prepare them?

Science is the tool used to satisfy the curiosity of a person who wants to look beyond the physical manifestations of objects and phenomena and, one glance at a time, to recognize and understand inter-relationships which those who are uninterested or uninformed will never comprehend. Science poses questions and the mere process of formulating those leads to a measure of insight even if they remain unanswered. Good, in-depth answers elicit new questions, which in turn lead to further insight and recognition that can intensify the *haiku* moment. Questions may relate to colours, shapes, and patterns and their transformation in time and space. Why is the flesh of salmon and tuna reddish, while that of flatfish is white? What happens when rice is cooked and fermented? Why are some teas bitter and feel harsh on the tongue while others have a smooth, well-rounded taste? How can one preserve vegetables and fruits so that they retain their flavour?

Here we are concerned with chemistry and the chemical reactions that take place between ingredients, with the physical properties of raw ingredients and tools, and with the biology of those living organisms which we use for food. One can easily prepare sushi

珍談 **AN OLD JAPANESE SAYING** has it that when you taste something delicious for the first time, especially the first fresh produce of the season, your life will be prolonged by 75 days. So, if you have never tasted sushi …

and enjoy the *haiku* moment without knowing the answer to these sorts of question, to say nothing of not even posing them in the first place. But I maintain that asking such questions will serve to sharpen the senses of intellectually curious individuals and that gaining knowledge of the science underlying the ingredients and the techniques employed in their preparation can greatly enhance the overall sushi experience.

DEMOCRACY IN THE SUSHI BAR

At a conventional restaurant you order a meal based on a written menu, you communicate your wishes to a server and then wait, possibly for a long time, until the food magically appears through a door which shields the kitchen and its activities from your sight. This has a completely different feel from what happens when you consume a meal at a sushi bar, probably the most democratic incarnation of a restaurant imaginable. Here you are a player in a dynamic process, over which you can exercise some influence.

It is the ultimate slow food experience, in which you participate in the process in its entirety. You can place your order directly with the chef. You see the fresh ingredients and are able to evaluate their quality

Kibune Sushi in Vancouver, Canada.

and follow along with their preparation. You can observe the way the meal is presented, which to top it all off is served by the chef himself. There is more – you can even change your order in mid-stream and ask the chef questions while he works. Simply put, there is no culinary culture that can offer an experiential chamber to rival the sushi bar.

The Japanese good luck cat, *Manekineko*, welcomes guests to the sushi bar.

WABI IS ALSO AN EXPRESSION used in Japanese political history. During the period when Edo (Tokyo) and Osaka overtook Kyoto in power and wealth, the latter made a virtue of necessity. The city of Kyoto became the embodiment of *wabi*, where the ideal of beauty and balance was to be found in the humble, the poor, and the worn. Simplicity was perfected.

By knowing a few rules you can transport this venue to your own kitchen. That does not necessarily mean that it is easy to prepare good sushi. This requires practice, patience, and an approach to the raw ingredients and methods of preparation that displays respect and a considerable degree of humility.

Practice is a good master. This is one of the reasons why it is said in Japan that it takes seventeen years to become a fully-trained sushi chef.

For the first couple of years, the apprentice washes dishes and fetches and carries for the chef. Next, a length of time is devoted to getting to know the various ingredients and how to choose fish at the market and clean them. Then the trainee has to learn to use the sushi knife and cut up fish. A period of intensive study follows to achieve perfection in the techniques of preparation and styles of presentation, although total perfection is neither possible nor necessary. This last point is part of my fascination with sushi – there is always room for improvement. And it is important to make something that is good. According to an ancient *Zen* saying, the person who makes something which is not good is worse than a thief who merely redistributes that which is good.

One can gain the impression that sushi is a terribly sophisticated cuisine, which in a way it is not. Sushi reaches sublime heights through the use of fresh ingredients "from the sea and from the mountains". But its presentation and the enjoyment of the meal can be completely down to earth or up in the clouds, according to how one looks at it. It is *wabi sabi*.

WABI SABI

Wabi sabi is a Japanese aesthetical-philosophical approach to finding beauty and meaning in nature. To the Western mind it often seems vague and difficult to grasp, but that very lack of precision is integral to the concept. *Wabi* signifies an inner quality that can be attached to a person, an animate or an inanimate object, and is characterized by modesty, solitude, sadness, simplicity, or stillness, building on the harmony found in all things in their natural state. *Sabi* stands for the outward traces left behind by usage and the passage of time, perceived as imperfection, insignificance, perishability, and wear. It embraces the melancholic beauty to be found in old, worn, and dilapidated things. *Wabi sabi* is rooted in the *Zen* idea that these outward manifestations of the ephemeral nature of life and things are the result of an eternal cycle of growth, decay, and death – the evolution of nothing into something and back again into nothing. An understanding of this process encourages one to focus on the fleeting beauty of the ephemeral – to reach for the wild strawberry as did the monk who was caught between two hungry tigers.

Wabi sabi permeates Japanese culture and way of thinking and has had a determining influence on a number of aesthetic expressions and art forms. The composition of *haiku* is one example. Others include the tea ceremony (*chanoyu*), flower arrangement (*ikebana*), landscaping, *Nō*-theatre, and *raku* ceramics. *Ikebana* is based on a triangular design in a simple container; gardening strives to form a whole landscape with a few stones; and the writer of *haiku* tries to express an aspect of the entire universe using just a few sound symbols.

One encounters *wabi sabi* at the sushi bar in the way the room is arranged, in the humility and attentiveness shown by both guest and chef to each other, in the manner in which the food is prepared and presented, and in the utensils which are used during the meal. The most perfect sushi might be served to you on an old, worn wooden plank. Your chopsticks could rest on a rough, irregular stone. You drink tea from a rustic *raku* cup with crackled glaze, which might even be cracked or chipped. The elegant tray on which the *sashimi* is placed might be irregular with, for example, a flaw in the glaze or an uneven rim. All these things are *sabi*. When the chef gives the guest his full attention and does not look down on him or her because of a possible deficiency of sushi expertise it is *wabi*. When

学術 *IKEBANA* ('the way of flowers') was originally a Japanese temple ritual in which flower arrangement was carried out as a meditative art form through which one could cleanse the soul and find harmony and balance. It was practiced only by men and *ikebana* was a component of the education of every *samurai*. The conventional Western approach to floral design focuses on the actual blossoms and their colour. *Ikebana* accords much greater prominence to linearity and simplicity, also emphasizing stems and leaves and the receptacle in which the arrangement is placed.

学術 *RAKU* MEANS something akin to 'pure pleasure' and denotes a distinctly Japanese type of ceramic. It developed in the 13th Century as a reaction to the delicate and overly decorated porcelain which came from China. With its high content of sand and finely ground fired clay, *raku* can withstand being removed quickly from the kiln and placed in cold air or sawdust. The glaze has to be simple and the bottom of a bowl or cup must remain unglazed, as it is said that this is where the soul emerges.

学術 NO GRAVY and preferably vegetarian. One of the secrets underlying the generally healthy, classical Japanese diet is the absence of gravy and rich sauces. Among other reasons, this is because food must be prepared in such a way that it can be eaten with chopsticks. Moreover, Buddhist precepts have periodically prohibited the consumption of milk products and of meats, sometimes even of fish. The traditional food in Japanese Buddhist temples, *shōjin ryōri*, was composed of few ingredients and was strictly vegetarian. It is still served to visitors at a number of temples.

the guest shows respect for the chef's work and appreciates the food for its taste and appearance it is also *wabi*.

When you develop an eye for it, you can see *wabi sabi* in all things, from the grandest to the most lowly. The sushi bar is a good place to start.

SUSHI AND WELLNESS – A LONG AND HEALTHY LIFE

Sushi and the traditional Japanese diet are profitable fields of study for those who want to learn something about how to achieve a long and healthy life.

A Japanese adult eats about 70 kilograms of fish a year, and thereby derives a much greater proportion of his or her proteins and fats from fish than does the average person living in the West. Even though the Japanese constitute only about 2.5% of the world's population, they eat 20% of the global fish catch and, furthermore, they eat about half of it raw. The average Japanese person presently has a life expectancy which is several years greater than that of Europeans and others who follow a Western lifestyle. It is easy to infer that a diet which includes sushi, among other foods, contributes to Japan's favourable position in this comparison, but it is equally obvious that many other factors can also play a role.

In the course of the 20th Century, Western countries have experienced an immense increase in the incidence of chronic, non-infectious diseases that are by-products of their physical and societal environments. Prominent among them are cancer and cardiovascular disease. Rare at the beginning of the last century, cardiovascular disease has in the course of the intervening years become the leading cause of death in the Western world. The mortalities attributed to this disease now exceed the total number of deaths caused by all infectious diseases put together. In the last few decades, there has also been an increase in the incidence of obesity, type 2 diabetes, high blood pressure, and fetal illness. Furthermore, it would appear that the incidence of mental illness, particularly among young people, is now growing at the same rate as cardiovascular disease did earlier.

It is generally thought that this shift in the disease patterns experienced in the West cannot be attributed to genetic changes. The

genetic inheritance of large populations is simply not transformed that quickly. In addition, the human genome is comprised of the surprisingly small number of about 25,000 genes. This hardly provides sufficient scope for variations that would enable changes in disease patterns of this magnitude to take place.

On the other hand, dietary changes can help to explain why deaths are chiefly linked to certain chronic illnesses. Food intake can alter gene expression and lead to a shift in the burden of disease in the course of a relatively short period of time. Those diseases associated with diet are referred to by the umbrella term the *metabolic syndrome*. Strictly speaking, the definition of this syndrome should also encompass a number of mental illnesses. Regrettably, public campaigns and programs to combat the metabolic syndrome have not been very effective. How can this trend be reversed?

Studying the combinations of foods eaten by groups of people who do not yet suffer very much from the metabolic syndrome can serve as a great inspiration. The emphasis on fish in the traditional Icelandic and Japanese cuisines and the olive-oil based Mediterranean diet are known to result in a low incidence of heart attacks and a long life expectancy. But the details of what effect the different elements of the diet have on each other are not known and it is likely that many factors are at play. In the meanwhile, it can be observed that those societies that begin to adopt a Western diet – some elements of which are fibre deficiency, large amounts of meat, overcooked and deep fried fish, a paucity of vegetables, and an excess of carbohydrates – increasingly are also manifesting disease patterns that are typical of the West. For example, if Icelanders start to eat less fish or Japanese to consume more fast-food hamburger combos, it can be expected that in the future these populations will collectively start to suffer from the lifestyle diseases that are now so common in many Western countries.

Fats *per se* play a vital role in human nutrition, but the types of fats consumed and how they are handled are also of key importance. A diet that is high in saturated fats, or in unsaturated trans fatty acids which are produced when plant oils are solidified to make margarine, is closely linked to the increase in cardiovascular disease. There is no doubt that the consumption of certain types of unsaturated fats is a determining factor in maintaining good health, both physical and mental. Superunsaturated fatty acids from oily fish, the so-called omega-3 fatty acids, contribute to lowering the risk of cardiovascular disease, reducing the cholesterol count, and decreasing the risk of cancer.

This is where sushi enters the picture. It is well known from investigations into mortality amongst Inuit and Japanese that their low incidence of cardiovascular disease is linked to a high consumption of fish.

Another significant group of unsaturated fats are the so-called omega-6 fatty acids, which are common in soybeans, also a major component of the Japanese diet. Omega-6 fats contribute to the production of certain eicosanoids, the vital hormones which regulate blood flow and strengthen the immune system. It is important to balance the intake of omega-3 and omega-6 fatty acids.

Scientific research has shown that the increase in psychiatric disorders can be attributed to an imbalance between omega-3 and omega-6 fatty acids in the diet. The present state of knowledge in these matters is still rather limited and insufficient to lead to concrete recommendations which go beyond general nutritional advice. But there are indications that just as cardiovascular disease was the Achilles heel in Western societies in the 20th Century, the central nervous system – particularly the brain – will be the next area of vulnerability.

It is a fact that sushi and fresh raw, preferably oily, fish together with seaweed can contribute to a balanced diet in which omega-3 fatty acids are well represented. In addition, sushi is low in calories and pleasantly filling. Soybeans and seaweed are rich in many important minerals and antioxidants which help to reduce the damage done to the machinery of the cells in the body.

Even if only a little is known about what can increase the possibility of having a long and healthy life, it is a virtual certainty that a moderate intake of calories and a diet which contains good, healthy fats are major factors in retarding the aging process.

All these indicators lead to the conclusion that sushi is a healthy food.

What is sushi?

The origins of sushi are rooted in fish preservation. Fresh fish starts to decompose quickly, but in earlier times there were no effective means of cooling or freezing it. The only possible way to prevent fish from spoiling in transit or in the warehouse was first to cure it in salt and then to ferment it. It was found that allowing the fish to ferment together with cooked rice resulted in an interesting taste and a pleasing texture. Sushi had been invented.

A BIT OF SUSHI HISTORY

Sushi is mentioned for the first time in a Chinese dictionary supposedly from the 4th Century, in this instance referring to salted fish that had been placed in cooked or steamed rice, which caused it to undergo a fermentation process.

It is not at all certain whether sushi was actually invented in China. It is thought that sushi was introduced to Japan in the 9th Century, but its origins are lost in the realms of the unknown. In this period, it was still common practice to drink milk and eat meat, but as Buddhism spread and the eating of meat was prohibited, people increasingly turned to the consumption of fish. It became necessary to devise new ways to store and prepare fish. This paved the way for the development of the Japanese sushi culture.

The combination of cooked rice and fermented, salted fish is called *nare*-zushi, which means aged sushi. The most widely known form of early *nare*-zushi is known as *funa*-zushi and was first prepared in the Shiga prefecture in Japan over a thousand years ago. *Funa* is a type of golden carp, which is common in Lake Biwa close to Kyoto. The carp was caught and salted at the start of summer. It was then prepared by being soaked in water to remove some of the salt, placed in a layer of cooked rice under pressure, and fermented for half a year or longer. After such a long fermentation, presumably only the fish was eaten and the rice was discarded.

Funa-zushi is also known from other countries in South-East Asia, among them Korea. In the 15th and 16th Centuries, a shorter fermentation period was introduced, typically of one month. The result was called *nama*-zushi (raw sushi) and, in this case, the rice was also eaten.

Modern sushi is related to *nama*-zushi, in which the fermentation of the cooked rice is hastened by the addition of rice vinegar. Production of rice vinegar grew rapidly in Japan at the start of what is known as the Edo period (1603-1867). The shogunate moved from Kyoto to Edo (Tokyo) and the production of rice, and with it the ancillary production of rice vinegar, skyrocketed. By adding rice vinegar to the cooked rice it was possible to shorten the fermentation period, but the process still took place under pressure. This type of sushi, which goes by the name of *haya*-zushi, is prepared in

Sushi. In Japanese the expression 鮨 means 'preserved fish' or 'fermented fish in rice and salt'. The Chinese *kanji* symbol on the left stands for fish. The two symbols on the right mean something like 'good taste'. Another common expression for sushi is 寿司.

学術 IT IS SAID THAT sushi was invented by an old Japanese couple who, out of the goodness of their hearts, had put out some leftover rice for a sea eagle that had built a nest close to their house. Later, when they discovered a fish in the nest, they took it home and ate it. They found that the fermented decaying rice had imparted a good taste to it – so they ate the fish and discarded the rice.

珍談 Iᴛ ɪs sᴀɪᴅ ᴛʜᴀᴛ sushi based on vinegared rice was invented in Edo (Tokyo) because the inhabitants, who were notorious for their lack of patience, did not want to wait for fermentation slowly to run its course. But it took a long time before modern *nigiri*-zushi spread from Tokyo to other parts of the country.

A sushi kiosk in Edo. Woodblock print by Utagawa (Ando) Hiroshige (1797-1858).

Nigiri-zushi is probably the best known type of *nare*-zushi: a simple, well-formed and slightly elongated ball of rice topped with a piece of fish.

the course of a 24-hour period and must be consumed immediately thereafter. The invention of *haya*-zushi is attributed to the Japanese medical doctor Matsumoto Yoshiichi who discovered that rice vinegar tenderized the fish and gave the rice a pleasant taste.

In the mid-1700's, the fermentation period was shortened to just a couple of hours with the introduction of *hako*-zushi, still made as a special form of sushi. Because it is prepared so quickly, it does not really involve fermentation *per se*. *Hako*-zushi is prepared by placing a layer of vinegared cooked rice together with filleted fish in a small wooden box which compresses the rice. To serve, the resulting block of fish and rice is cut into slices.

Tradition has it that in the 1820's Hanaya Yohei (1799-1858) from Edo invented or elaborated the modern form of sushi, which is called *nigiri*-zushi. It consists of a simple ball of rice, shaped by hand, with a piece of fish placed on top of it. The rice used is freshly cooked, after which rice vinegar and salt are added. This can be considered 'speed fermentation' of only a few minutes duration. The fish is completely fresh and does not have time to be preserved by contact with the vinegared rice and, in contrast to the original *nare*-zushi, both fish and rice are eaten immediately after preparation. Pressure, apart from that applied by the hands to shape the rice ball and attach the piece of fish, is no longer employed. In this way, sushi was transformed into an early version of fast food. *Nigiri*-zushi has come to symbolize sushi as it is now known around the world.

There is little doubt that *nigiri*-zushi was intended for ordinary people who in the course of a busy day could, without much fuss and bother, grab a couple of pieces of sushi at one of the many outdoor kiosks found all over Edo in the 19th Century. This was much like the current practice of casually picking up a ready-made sandwich or a hot dog. After the great earthquake of 1923, the sushi-stands in Tokyo moved indoors and were transformed into proper bars or restaurants.

The evolution of sushi did not stop at *nigiri*-zushi. In Japan it has continued to develop both as an everyday food and as a form of culinary art with a wealth of local variations. An example of the latter is *oshi*-zushi, prepared under pressure and usually with mackerel, which is characteristic of the Osaka area (Kansai). *Oshi*-zushi can

One of the earliest illustrations of *nigiri*-zushi comes from the famous Tokyo sushi shop of Hanaya Yohei (1799-1858), which closed only in the 1930's. The illustration was adapted by an unknown artist from pictures drawn by Kawabata Gyokusho (1842-1913).

学
術
Edomae-zushi or *nigiri*-zushi? *Nigiri*-zushi is also known as *Edomae*-zushi. *Edomae* refers to the small bay in Edo in front of the old palace that stood on the same site as the present-day imperial precinct in Tokyo. Fresh fish and shellfish caught in the bay were used locally to make sushi, which was known as *Edomae*-zushi. It has, however, been many years since these waters have been a source of seafood. Now the expression *Edomae*-zushi is employed as a synonym for high quality *nigiri*-zushi.

be considered a modern version of *hako*-zushi. Another example is *sugata*-zushi, which consists of a whole fish cut open and stuffed with sushi rice and then presented in its original shape. Different regions of Japan utilize different fish for this dish.

Sushi rolled in sheets of seaweed, known as *maki*-zushi, seems to have been invented before *nigiri*-zushi, possibly in the mid-1700's or even earlier. The rolls were pressed together using a simple bamboo mat, a practice which continues to the present day.

In the course of the 1970's, interest in sushi spread to North America. Since then a succession of the best Japanese sushi chefs have opened restaurants abroad, particularly in the United States. During the past decade, sushi has become a global phenomenon, given a significant helping hand by a heightened interest in Asian cuisine and culture and the intense focus on a healthy diet.

To give but one example: California has become a sort of Mecca for modern sushi-culture, and the Californian sushi chefs have introduced numerous exciting new elements, experimenting with vegetarian sushi dishes, using spices not traditional in Japan, and local fish and shellfish.

Rolled sushi, a California roll.

珍
談
IT IS A DUTCH CUSTOM to eat the small herring from the North Atlantic and the North Sea, which are caught in the first weeks of the herring season at the end of May, as 'nieuwe haring' (new herring). The fish have been lightly salted, and frozen at sea, leaving them nearly raw, and are consumed immediately after thawing and the final cleaning, when the heads are cut off and the innards removed. To eat one, take the fish by the tail, lean your head back, and let the fish melt in your mouth. In accordance with tradition, you can also dip the herring in raw onion before you dispatch it.

Special sushi rolls (*maki*-zushi) have taken on whole new forms in Californian sushi bars, in fact, one of the now standard rolls – with avocado and crabmeat – is even called a California roll. In addition to this, a great deal of experimentation involving sushi and other Japanese specialities takes place under the umbrella of fusion cuisine.

It is interesting to note that, despite the fact that the original sushi, i.e., fermentation of rice together with fish, was found in many parts of Southeast Asia and probably also in Polynesia, it is only in Japan that this type of food has been refined in the course of the past thousand years. Perhaps the Koreans would take issue with this conclusion.

SUSHI IS VINEGARED RICE WITH SOMETHING ON TOP (*TANE*) OR INSIDE (*GU*)

The main principle in every form of modern sushi is a combination of vinegared cooked rice either with something placed on top (*tane*, *neta*) or with a filling (*gu*). The same ingredients can play a role both as *tane* and as *gu*. That which is *tane* in *nigiri*-zushi becomes *gu* when it is used to stuff a *maki*-roll.

There are four classical types of *tane*. *Akami* are red or dark *tane*, for example, tuna and salmon. *Shiromi* are white *tane*, such as flatfish with white muscle flesh. *Hikari-mono* are shiny *tane*, typically mackerel and herring with the skin left on. *Nimono-dane* are *tane* which have been cooked or simmered, often octopus, some bivalves, and eel. *Hokanomono* are *tane* which are not encompassed by the above categories, for example, shrimp, roe, and sea urchin.

In rolled sushi, *gu* is the designation for everything other than the rice – fish, omelette, *tofu*, spinach, cucumber, sprouts, other vegetables, crabmeat, mushrooms, green *shiso*, pickled radish (*takuanzuke*), sesame seeds, and so on. Before they can be used, some types of *tane* and *gu* have to be prepared, either by cooking, simmering, salting, or marinating, in order to make them edible or to bring out the right flavours. Others are eaten completely raw, in some cases after having been frozen for a period of time.

The question is: how fresh must raw *tane* and *gu* be? In the case of sushi fish, one speaks of *ikijime*, which is fish that is consumed right away after it has died and before *rigor mortis* has set in. In order to limit the struggling of the fish, it is usually killed in iced saltwater. Sometimes this takes place right behind the sushi bar where the fresh fish is then cut up and immediately made into *tane* by the chef. This is often a variety of white fish, as the flesh of this type of fish has a firm texture which is not made more tender by natural decomposition. Other fish need to be 'ripened' before they are eaten and these are referred to as *nojime*. They have gone through *rigor mortis* and have been frozen for a period of time. As a consequence, their taste and texture have changed due to natural decomposition. *Nojime* intended for sushi have to be eaten as soon as they have been defrosted. Red fish, like salmon and tuna, are commonly treated in this way.

Sushi in the famous Matsuno sushi store in Edo. Woodblock print by Utagawa Toyokuni III (1786-1864).

A 16th Century Nordic fishmonger. Several different types of preserved fish are being sold: dried fish, smoked fish, and salted fish in barrels. Based on Olaus Magnus: *Historia de Gentibus Septentrionalibus* (1555).

'Kalakukko', traditional Finnish fish bread, in which the fish is baked in dough, thereby preserving it in bread.

THE NORDIC ANSWER TO SUSHI

There is a well-known Nordic variant on sushi, namely, an age-old tradition of preserving fish by combining them with other foodstuffs which contain a large proportion of carbohydrates. An example of this is 'kalakukko', fish baked in bread dough, still prepared in some rural districts of Finland.

In the Middle Ages it was common practice to conserve salmon or herring by lightly salting them and sometimes adding flour or barley malt, wrapping them in bark (often from birch trees), and then burying them in the ground. In the cool soil, the fish underwent a fermentation process with the help of lactic acid bacteria, which, together with the *enzymes* contained in the fish itself, preserved the fish and transformed it into 'sursild' (sour herring) or 'gravlaks' (salmon buried in the ground), both of which have a sour and sharp smell and taste.

In the 1700's this Nordic sushi technique became more sophisticated, introducing a way to use less salt. The fish is placed under pressure in a cool environment to absorb the salt and a little sugar, possibly with some seasonings such as dill or peppercorns. Although the fish is not completely preserved by this process, it will keep for a few days after it has been cured. This method of preparation is now also used for other fish, for example, 'siika'(Baltic whitefish).

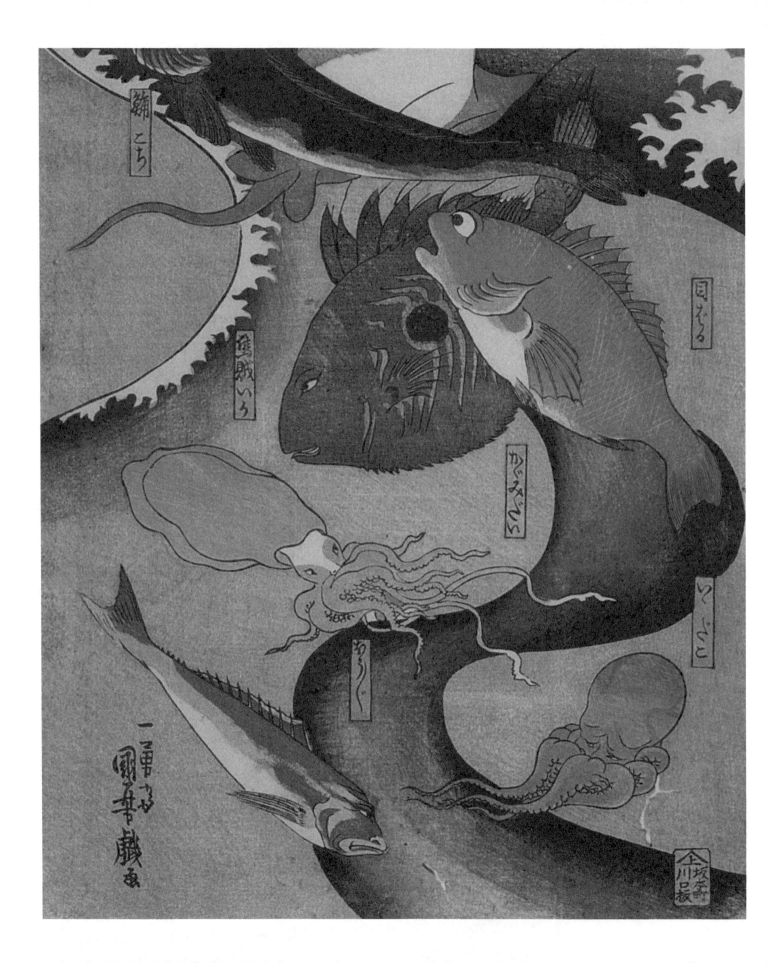

SUSHI VARIATIONS

Handshaped sushi: *Nigiri*-zushi

Rolled sushi: *Uramaki* (inside-out rolls)

Rolled sushi: *Hosomaki* (slender rolls)

Battleship sushi: *Gunkan-maki*

Rolled sushi: *Futomaki* (thick rolls)

Pressed sushi: *Oshi*-zushi

Handrolled sushi: *Temaki*-zushi

Scattered sushi: *Chirashi*-zushi

LIFE, FOOD & MOLECULES

If you understand, things are just as they are ...
If you do not understand, things are just as they are ...
Zen saying

THE MOLECULES OF LIFE

If we exclude water, the greatest part of our food and drink is of organic origin and is derived from plants or animals, which themselves are more than half made up of water while in living form. Water is, therefore, the most important single substance for life and its sustenance. Organic material from plants and animals is classified into four different categories: proteins, fats, carbohydrates, and nucleic acids. The first three categories supply the most vital molecular components in our food, also sushi.

CELLS AND MOLECULES

All living organisms are composed of *cells*. Some of them are uni-cellular, for example, bacteria and yeast. Others are multicellular, containing from a few hundred to billions of cells. A human being is made up of about 100,000 billion cells.

Despite the fact that individual organisms may look very different and may exist under widely varying conditions, they have a com-monality – all are made from the same molecular building blocks and they are controlled by the same types of chemical reactions. All cells are built up from small organic molecules, which all contain carbon. These small molecules can be subdivided into four classes: *monosaccharides*, *amino acids*, *fatty acids*, and *nucleotides*.

The four types of molecular building blocks are able, on their own, to combine with each other or with molecules from the other classes. In this manner they can create larger molecules, the so-called *macromolecules*, or *macromolecular assemblies*.

Macromolecules, in turn, are also divided into four categories: *carbohydrates* (*polysaccharides*), *proteins*, *fats*, and *nucleic acids*.

Macromolecules from the first three categories are abundant in nearly all forms of food and each of them plays an essential role in providing sound nutrition.

学術 WATER, LIFE, AND FOOD. All life, as we know it, is based on the premise that *water* is present. Hence, when one looks for signs of life beyond the Earth, one first tries to detect the presence of water or traces of it.

Since all of our fresh food contains water, it is important to understand its physical and chemical properties in order to handle and prepare food properly.

Schematic representation of the structure of an animal cell. The *cell* is surrounded by a *cell membrane*. Inside the cell are found the nucleus, which contains genetic material (*genes*), and the various organelles. Plant cells resemble those of animals, but have an extra rigid cell wall composed of *carbohydrates*.

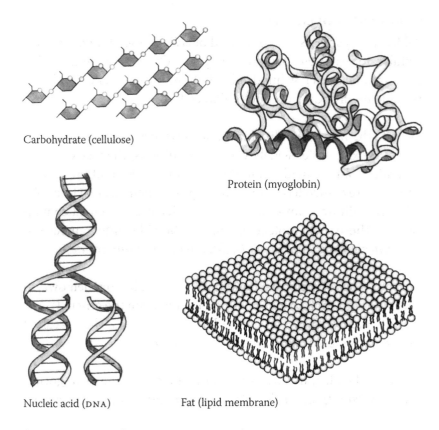

Carbohydrate (cellulose)

Protein (myoglobin)

Nucleic acid (DNA)

Fat (lipid membrane)

SACCHARIDES (CARBOHYDRATES)

Nature avails herself of many different *monosaccharides*, for example, *glucose*, to form *polysaccharides*, which are also referred to as *carbohydrates*. The simplest, mono- and disaccharides, are sweet and we know them as *sugars*. Polysaccharides are long chains of saccharides, which in many cases are crosslinked in all directions. This helps to give polysaccharides some special properties that are exploited to strengthen cell walls and to create strong fibres. *Cellulose* is an example of a robust network of polysaccharides.

Polysaccharides are important as fuel for the cells. An example is *glycogen*, a polysaccharide composed of *glucose*. *Starch* is another well-known polysaccharide; it is a mixture of the polysaccharides *amylose* and *amylopectin*.

AMINO ACIDS AND PROTEINS

Nature makes use of 20 different *amino acids* in order to make *proteins*. A protein is a chain of amino acids; this is why it is con-

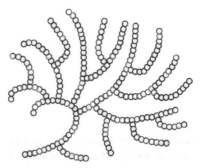

Glycogen is a branched *polysaccharide* molecule, which consists of *glucose*-units (shown as blue spheres). *Glycogen* stored in the liver and the white musculature of fish and shellfish is the energy depot of the organism.

sidered a *polymer*, more specifically a *polyamide*. Proteins can be quite short, composed of only a few amino acids. An example of this is melittin, the principal active component of bee venom, which contains a mere 26 amino acids. Others are much longer. There are up to a thousand amino acids in *gluten* proteins found in wheat. In the protein *myoglobin*, the substance that is responsible for the red colour of the muscles in animals and certain fish, there are 153 amino acids.

Of the 20 amino acids there are nine, called the *essential amino acids*, which our bodies cannot produce and which we must obtain from our food intake.

In order for a protein to be able to fulfill its proper function, the long chain of amino acids of which it is made up must fold itself together in exactly the right way. The protein's structure and function can be damaged by external physical effects such as heating or cooling it or putting it in contact with acid. The result is a *denatured* protein. This phenomenon is well known to everyone from the example of a cooked egg, in which the proteins have unfolded and clumped themselves together into a solid mass. The same thing happens when meat or fish are cooked or roasted; the muscle proteins are denatured and become firmer and tougher.

Enzymes are a special class of proteins, which catalyze chemical reactions, that is to say, they speed them up. Our cells are full of enzymes, which assist in breaking down food in the stomach and the intestines and in building new cells. Enzymes found in foodstuffs, for example, those in the muscles of fish and shellfish, are particularly aggressive and quickly catabolize the muscle tissue unless it is cooked, frozen, or marinated in an acid (usually vinegar). These processes denature the enzymes, or at least significantly slow down the speed at which they work.

Free amino acids, which are not bound in proteins, play an important role in the taste of many of the ingredients used to prepare sushi. Examples are the amino acid *glycine*, which imparts a slightly sweet flavour to crabmeat, and *glutamic acid*, which gives seaweed its distinctive taste. Molluscs such as octopuses and bivalves derive their taste from alanine, proline, and arginine.

学術 **ESSENTIAL AMINO ACIDS.** The nine essential amino acids are valine, leucine, lysine, histidine, isoleucine, methionine, phenylalanine, threonine, and tryptophan.

Denaturing of a *protein* or an *enzyme*, for example, by heating it or exposing it to acid. The individual blue and red spheres represent *amino acids*, which can be dissolved in water or oil, respectively. In its folded state (to the left), where the red amino acids are shielded from contact with the water in which it finds itself, the molecule is active, whereas it has become inactive in its unfolded, denatured state (to the right).

学術 **ENZYMES** are a special class of proteins. *Proteins* are the work horses of an organism, carrying out many different tasks. Some transport molecules while others relay signals, for example, those originating in the nerves. Other proteins are able to differentiate chemical substances by taste or smell and still others build up the structures of the *cells* and give them mechanical strength.

Enzymes mediate most of the chemical processes that are crucial for such vital functions as energy conversion and the breaking down of materials and the rebuilding of cells and molecules in the body.

FATS AND OILS

Fats and *oils* constitute a large, varied body of substances which share the characteristic that they cannot be mixed with water. There is no particular difference between an oil and a fat. Normally one refers to substances such as wax, lard, and butter as fats because they are solids at room temperature, whereas those which are liquids, like olive or fish oil, are called oils. But we all know that butter melts when it is heated and that fish oil becomes solid when it is frozen. Oils are really just melted fats. The melting point of a fat has major implications both for its taste and nutritional value. When we talk about saturated and unsaturated fats, there is often a reference to a difference in melting point. Saturated fats melt at higher temperatures than the unsaturated ones. We are aware of this distinction from hard butter produced from animal fats, primarily saturated, and soft margarine made from vegetable oils, partially unsaturated.

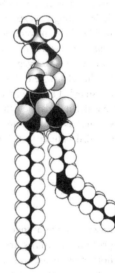

Fats can be solid, e.g., hard butter, or liquid, e.g., olive oil.

A *lipid molecule* consists of a water soluble head and two fat soluble tails of *hydrocarbon chains* which, in the illustration above, are stearic acid (left tail) and *oleic acid* (right tail). Stearic acid found, for example, in sunflower seeds, is a completely saturated fatty acid and oleic acid found, for example, in olive oil, is a monounsaturated fatty acid with a single double bond at the point where the chain bends.

Lipids are an important type of natural fat. They are molecules made up of *fatty acids* and an assortment of other components, for example, *amino acids* and *saccharides*. The walls or *membranes* that enclose all cells are built up with lipids. In contrast to saccharides, amino acids, and nucleotides, lipids do not form *polymers*. Instead they form *macromolecular assemblies* such as a bilayer *membrane* or a *liposome*.

The reason for this distinctive phenomenon is that lipids have mixed feelings about water. On one end of a lipid molecule there is a head, which dissolves in water, but in the other end are two

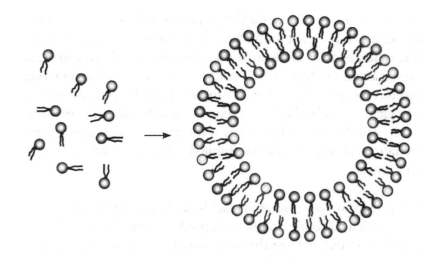

Self-organization of *lipid molecules* in water, leading to the spontaneous formation of a *membrane*. Lipid membranes are a double layer of lipid molecules which form a sphere around each other to make what is called a *liposome*. Water is found on both sides of the membrane.

fatty tails derived from fatty acids, which are immiscible in it. The compromise has been to form a bilayer, known as a *lipid membrane*. Within the membrane the fat soluble parts of the lipids are, for all practical purposes, shielded from contact with water. Hence, lipids are described as being *amphiphilic*, which is to say that they have a love-hate relationship with water.

Embedded in the *membranes* there is an assortment of *proteins* and *enzymes* that govern most of the functions of cells, such as communication with the immediate environment and transport of materials in and out of the cells.

An example of this is that all nerve cells are surrounded by membranes that have special proteins embedded in them which act as tiny channels for the passage of important *ions*, e. g. sodium and potassium. These membranes also contain a variety of *receptors*, including those which mediate sensations of smell and taste. Other receptors come into play when the nerve cells are affected by different medications.

Recent research has shown that different fats, that is to say, *lipids*, have a different effect on the function of the cell membrane in which they are found.

学術 **DOES IT LOVE WATER, YES OR NO?** Materials are often categorized according to their relationship with water, that is to say, whether they love water (*hydrophilic*) or hate it (*hydrophobic*). Substances which are hydrophilic can easily be dissolved in water – think of household salt and sugar. Oils and other fats are hydrophobic and cannot form a mixture with water. Many frying pans on the market today are coated with a layer of polymers, which are strongly hydrophobic, in order to make the pans non-stick.

Those materials that have a love-hate relationship with water are refered to as *amphiphilic*. Both soaps and fats (*lipids*) are amphiphilic.

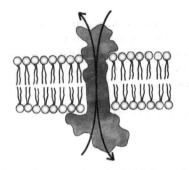

A *membrane* consists of a double layer of *lipid molecules*. In the illustration above, the membrane is shown with an embedded *protein*, which can function as a *receptor*, for example, by recognizing molecules which convey smell and taste or by acting as a channel to allow *ions* or other molecules to pass through the membrane.

学
術 CHOLESTEROL is a special fat, which is a major component of our *cell membranes* and which, in addition, is the raw material for the production of such important substances as male and female hormones, bile salts, and *vitamin* D. The distribution and transport of *cholesterol* within the body is controlled by certain *lipoproteins*. If there is an imbalance between this transport system and the liver's ability to produce, as well as to break down, cholesterol, the possibility of developing hardening of the arteries is increased.

Of particular significance for the properties of fats and lipids is the degree of saturation of the *fatty acids* of which they are composed. A fatty acid consists of a long *hydrocarbon chain*, which is a string of carbon atoms to which hydrogen atoms are attached. Neighbouring carbon atoms in the chain are chemically bound together by either a single or a double bond. The degree of saturation is dependent on the number of double bonds – the more there are, the more the fatty acid is said to be unsaturated; when there are none, it is fully saturated.

Just as there are essential amino acids, there are also essential *fatty acids,* which our body cannot itself produce, and which we, therefore, have to obtain from the foods we eat. The two essential fatty acids in question are both polyunsaturated: *linoleic acid* has two double bonds and *alpha-linolenic acid* has three.

These two essential fatty acids are the starting point for the synthesis of two different families of important fatty acids, the so-called *omega-3* and *omega-6* fatty acids. Our bodies cannot easily convert fats from the one family into those of the other, so it is vital to obtain both types from our food. Major sources of omega-3 fatty acids are salt water fish and shellfish, seaweed, algae, meat, and egg yolks. Omega-6 fatty acids can most easily be obtained from plants, with sunflower oil, soybeans, and corn being among the more common ones.

Flax seeds are also an important source of *alpha-linolenic acid.* Research has shown that only about 5% of the alpha-linolenic acid in the diet is converted to the important, superunsaturated omega-3 fatty acids, DHA (*docosahexaenoic acid*) and EPA (*eicosapentaenoic acid*), which are found in fish and shellfish. The conversion can be further limited if there is a high proportion of saturated and omega-6 fatty acids in the diet. It is interesting that women, in particular during pregnancy, are slightly more able than men to derive these omega-3 fatty acids from flax seeds. In any case, it is more efficient to obtain one's omega-3 fats from fish than from flax seed.

Sushi is, therefore, an ideal source of essential fatty acids.

Saturated and unsaturated fats

Whether a *fat* is saturated or unsaturated depends on how many double bonds there are in those *fatty acids* of which it is composed. Polyunsaturated fats have more than one double bond. Unsaturated fats can be oxidized, which breaks down the double bonds, causing the fats to go rancid.

Saturated fat

has no double bonds – for example, palmitic acid derived from palm kernel oil.

Monounsaturated fat

has one double bond – for example, *oleic acid* found in olive oil.

Polyunsaturated fat

has more than one double bond – for example, there are two double bonds in *linoleic acid* derived from soybeans and there are three double bonds in *alpha-linolenic acid* extracted from flax seeds and seaweed.

Superunsaturated fat

has more than four double bonds – for example, there are six double bonds in DHA (*docosahexaenoic acid*) sourced from fish oils.

Omega-3 fats

are essential *fatty acids* which are based on *alpha-linolenic acid*. Two important examples are the superunsaturated fats, DHA (*docosahexaenoic acid*), which has six double bonds, and EPA (*eicosapentaenoic acid*), which has five double bonds. Fish and algae are rich sources of these omega-3 fatty acids, which play a vital role in the function of the nervous system. In addition, their many contributions to the maintenance of good health are well documented – they help to prevent circulatory problems, inflammation, cancer, and certain mental illnesses.

Fats in fish

are overwhelmingly superunsaturated fats. One reason is that the fat found in the muscles of fish must remain fluid at cool temperatures in the oceans. Superunsaturated fats have very low melting points.

珍談 IT IS SAID THAT some people think it is dangerous to eat *genes*, which of course is utter nonsense. Genes in the form of DNA are found in all cells and are, therefore, present in all the food that we consume. Once in our digestive tract, the genes from plants and animals are broken down into the four different *nucleotides* of which they are composed. These are the same nucleotides as the ones which make up human genes.

Our body then reconfigures these nucleotides to match exactly our own unique genes. Genes found in our food and, by extension, also those found in genetically modified products, cannot be mixed with our own genetic material.

NUCLEOTIDES, DNA, AND GENES

A *nucleotide* is composed of three parts: a base, a *sugar*, and one or more phosphate groups. There are five fundamental bases: adenine, guanine, cytosine, thymine, and uracil. The nucleotides polymerize and form a long chain, a *polynucleotide*. DNA (*deoxyribonucleic acid*) is a polynucleotide composed of the first four bases listed above. Normally it consists of two polynucleotide chains that are twisted around each other, forming the well-known double helix. RNA (*ribonucleic acid*), like DNA, is composed of four bases but in this case uracil replaces thymine and the sugars are different.

A single molecule of DNA can be very long. These molecules store our genetic information and are known collectively as the *genome*. Each of our cells contains a piece of DNA, which would be almost a meter long if stretched out.

Nucleic acids are not particularly important in relation to nutrition, but some of them help to bring out the *umami* taste found in very fresh fish. When the cells in the muscles of the fish need to produce energy, the fuel ATP (*adenosine triphosphate*) is broken down, producing IMP (*inosine monophosphate*) and GMP (*guanosine monophosphate*), both of which have *umami* taste. These substances start to disappear gradually as soon as the fish dies.

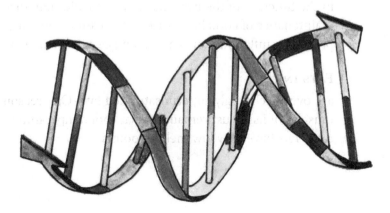

The DNA molecule, which is a double spiral, is the backbone of all genetic material.

Cholesterol and evolution

ABOUT A MOLECULE WITH A BAD REPUTATION

Many people think of fats and cholesterol as twin evils – they are a health hazard and are to be avoided at all costs. This has given birth to some truly bizarre and laughable practices in food labelling. As an example, the packaging of some cane sugar found in cafés declared it to be cholesterol free, even though it is obvious that it has never had even the remotest connection with fats. Nevertheless, many people realize that, despite its bad reputation, cholesterol plays a vital role in the body's ability to synthesize important hormones, among them sex hormones, vitamin D, and bile salts, which help to digest food in the stomach. But most are not aware that fats, especially cholesterol, are indispensable for cell function and that cholesterol is the most common fat molecule found in our cell membranes. How did this come about and what particular properties does this confer on the cells?

Chemically speaking, cholesterol is considered a higher sterol, that is to say, of the type used by animal life. There are parallel higher sterols, e.g., phytosterol, fucosterol, and ergosterol, used by plants, algae, and fungi, respectively. Healthy people cannot convert one variety of sterol into another. For this reason, the sterol content of foodstuffs derived from plants, algae, and fungi normally does not increase an individual's cholesterol reading.

As we shall see below, Nature has gone to great lengths to evolve higher sterols. For the sake of convenience, they will be referred to simply as cholesterol in what follows.

Most people know that Darwin's principles concerning evolution hold for both flora and fauna and that they are manifest at the molecular level in the genetic hereditary material of each species, the genome. It is less well known, however, that molecules other than DNA have been subjected to the laws of evolution. Cholesterol is still the unparalleled demonstration that this is the case.

Life arose on Earth about 3.8 billion years ago. At that time, those chemical substances were present that, under the conditions then current, made possible the formation of the first primitive cells. These early cells, which did not contain cholesterol and are called prokaryotic, are similar to the most primitive present-day bacteria. There is a reason why these prokaryotic cells did not have any cholesterol – it was not yet to be found.

The explanation for this came from the American biochemist and Nobel laureate Konrad Bloch, who proved that the biochemical synthesis of cholesterol is possible only in the presence of molecular oxygen. When life arose, there was effectively no free oxygen in the Earth's atmosphere, less than one part in ten billion. With the passage of time, those organisms that gave off oxygen as a waste product, especially the blue-green algae, evolved. The resulting accumulation of oxygen in the atmosphere can be characterized as a global environmental catastrophe because those forms of life that are oxygen intolerant had either to become extinct or to bury themselves in the oxygen deficient seabed. As the oxygen content of the atmosphere increased, new forms of life evolved that could incorporate it into their metabolic function, the process we call respiration. Up to that point, all living things were prokaryotic.

The formation of large quantities of molecular oxygen set new evolutionary forces in motion. New molecular building blocks could take shape and, in the course of the unfathomably long time span associated with evolution, there was a great likelihood that these molecules actually would be generated. If an organism can gain an advantage by utilizing them, it will do so and, as a consequence, be successful in the Darwinian selection process. The new type of molecule that was made possible was cholesterol and the new type of organism which turned out to be able to take advantage of this molecule is the progenitor of the higher forms of life, the eukaryotes, that is to say, plants, fungi, and animals. This viewpoint is supported by the fact that a plethora of eukaryotes teemed forth between 2.4 and 2.8 million years ago, at precisely the same point in time as the concentration of oxygen in the atmosphere was rising sharply.

So one can say that cholesterol removed a sort of bottleneck in the evolutionary process. This statement is supported by the fact that all eukaryotes contain large quantities of cholesterol in their cell membranes, whereas it is absent in all prokaryotic organisms.

An interesting observation is connected to this fact. Prokaryotic cells have only outer membranes, whereas the cellular structure of eukaryotes is more complex. Over and above their outer membranes, the eukaryotic cells also make use of membranes internally to encapsulate their nuclei and various organelles, among others the mitochondria. Mitochondria are the power houses of the cells, producing the energy they need for their varied functions. Interestingly, the membranes of mitochondria contain very little cholesterol. Why?

A possible explanation can be found in a hypothesis formulated by the American microbiologist Lynn Margulis, who has made herself into a spokesperson for what is known as the symbiosis theory.

According to this theory, mitochondria are former prokaryotes which early eukaryotic cells engulfed in order to exploit their ability to carry out respiratory functions, that is, to use oxygen to produce energy for the cell.

In return, these early prokaryotes were afforded the luxury of living in a more sheltered environment inside the eukaryotic cells. A beautiful example of symbiosis! These prokaryotes, now in the form of mitochondria, have preserved aspects of their original membrane and, consequently, contain very little cholesterol.

The question then arises: what evolutionary advantages can cholesterol impart that simpler sterols cannot?

Recent research has shown that it is cholesterol's ability to serve as a sort of anti-freeze in the membranes that might be the key to its success in the course of the evolution of the species. The cell membrane is composed of lipids with different degrees of saturation; the more saturated, the more viscous the membrane. It is important to have reasonably fluid membranes in order to support the different cell functions, but fluid membranes are less stable.

It turns out that cholesterol can save the day. It ensures that the membranes are sufficiently fluid while at the same time providing an appropriate degree of rigidity and stability. The biochemical precursor to cholesterol, called lanosterol, which is formed in the absence of oxygen, is unable to perform this important mechanical function.

SENSORY PERCEPTION

The sensory experiences that come with eating and drinking are a wonderful combination of seeing, tasting, and smelling, as well as feeling the texture, consistency, and temperature of food when we put it in our mouth. There are five types of taste. In addition to sweet, sour, bitter, and salty, there is a fifth taste, called *umami*, which is centrally important in Japanese cuisine, not least of all to sushi. *Umami* is especially associated with the taste of shellfish, smoked fish, and seaweed.

How does food taste?

Taste and smell can, in principle, be defined chemically, in the sense that one can isolate and characterize the chemical components that release a certain taste and smell. One can also determine the exact physiological and biochemical mechanisms involved in the sensory perception of one or more components of food and drink.

It quickly becomes more complicated when the sensory perception arises from a combination of chemical impressions. Given that there can be chemical reactions between the various ingredients in the food, between food and drink, and between the food and the chemicals found in the oral cavity itself, the description of a given sensory experience can become extremely complex and perhaps even specific to the individual in question.

In this connection, spices occupy an exceptional position as ingredients that not only add flavour to the food but also, in their own special and characteristic way, modify its taste. For example, a spice might bring out a subtle taste component in the food or mask another, possibly unwanted, one.

It can be very difficult to find words to express sensations of taste and smell which would make it possible to communicate these experiences to others. Frequently one resorts to saying that something tastes or smells more or less like something else which one supposes is already well known to the listener.

The transformation that impressions of taste and smell often undergo in the course of a meal adds another intricate dimension to the sensory picture. Sometimes there is a first sensation, a second impression, etc., and we even talk about an aftertaste.

Finally, a number of psychological factors also influence the senses: sensations of fullness, the intensity of the aesthetic experience associated with the visual impact of the food, the type of eating utensils and dishes used for the meal, the overall presentation and service, the arrangement of the physical surroundings, lighting, sound, and, of course, the company or lack thereof in which the meal is eaten.

An overarching theme of this book is that all the senses should be firing on all cylinders both in the preparation and consumption of the wide spectrum of items which I have included under the rubric of sushi. None of the above mentioned elements of the experience can be separated from each other without altering the experience itself. There is something in it for the eyes, the body, and the soul.

Biochemically there is a high degree of understanding of what taste and smell are, and of the interplay between them. A series of special molecules (*receptors*) which convey the sensory impression from the organ that perceives it to the brain have already been identified. Furthermore, the fundamental aspects of food chemistry, as well as those related to food preparation and conservation, are known in great detail. In many instances one has been able to characterize the so-called active or distinctive chemical components and discover how they can be transformed. This insight is frequently used in order artificially to impart a particular taste or smell to a food or a drink by the use of what are classified as additives.

Nevertheless, when one considers that any given drink or foodstuff can be composed of thousands of different chemical substances, of which only a very small number may have been identified and characterized, it is clear that our knowledge of taste and smell is very limited. Food chemistry is complex, and the same holds true for the chemistry underlying gastronomy. These relationships are thoroughly described in the wonderful book *On Food and Cooking: The Science and Lore of the Kitchen* by Harold McGee (2004).

Molecular gastronomy has in recent years been introduced as the designation for the science associated with the molecular properties of food ingredients and the ways in which handling and preparation transforms them. The two pre-eminent figures in this discipline are the late British physics professor Nicolas Kurti and the French chemistry professor Hervé This, both passionate advocates of applying scientific principles to the formerly primarily empirical culinary arts. One aspect of molecular gastronomy is to formulate molecular and quantitative explanations for taste and taste sensations, as well as how different sensory perceptions enter into a synergistic relationship with each other.

Umami – the fifth taste

There is a tradition going back to classical Greek literature that there are four types of taste – sweet, sour, salty, and bitter. In 1908 the Japanese chemist Kikunae Ikeda posited the existence of a fifth taste sensation. He dubbed it *umami*, which means something along the lines of 'delicious' or 'palatable'. The taste is due to the substance *MSG* (*monosodium glutamate*). Ikeda found that brown algae and edible kelp, for example, the Japanese *konbu*, are especially rich in MSG, which often precipitates as a deposit of crystalline powder on the surfaces of the dried seaweed fronds.

Umami taste is also closely linked to a number of substances that are based on nucleic acids, particularly *inosine monophosphate* (*IMP*) and *guanosine monophosphate* (*GMP*), which are produced when the biomolecule *ATP* is broken down in the cells in order to release the energy contained in it.

Konbu has for many centuries been used to add flavour to soups and other cooked dishes. In this sense, *umami* taste should perhaps have been described as a flavour intensifier rather than as a separate taste. But in 2001 it was shown that humans and other animals actually have a specific taste *receptor* for MSG. With this discovery it became clear that there is also a verifiable physiological basis for designating *umami* as the fifth primary taste, distinct from the initial four.

Since then a couple of other chemical substances which bring out the *umami* taste have been found in *shiitake* mushrooms, among other sources. The different substances seem to have a synergistic effect, in that a tiny amount of one can enhance the taste experience of another.

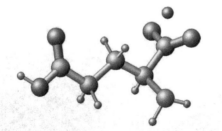

The molecule *MSG* (monosodium glutamate) is the sodium salt of the *animo acid, glutamic acid.* MSG, also known as 'the third spice' because it is the most widely used after salt and pepper, imparts the *umami* taste.

Taste

Since antiquity, Western culture has divided taste sensations into four types: sweet, sour, bitter, and salty. But in Asia, one has for many years allowed for a fifth taste, *umami*. *Umami* is important for the taste sensation of the food which is the subject of this book. It is sometimes described as a 'meaty' or 'brothy' taste, although this does not really encompass what is meant. *Umami* is best known from the taste of mushrooms and MSG, 'the third spice'.

The perception of taste presupposes that the substances that convey taste are first dissolved in liquid in the mouth and the throat. Sensation is localized on the tongue and partially in the oral cavity (particularly in children) in small sacs (papillae). Those at the front of the tongue contain the sacs known as the taste buds. In each of the taste buds, there are up to a hundred different specialized taste sensory cells arranged in an onion-shaped formation.

The way the sensory cells work is that a certain type of sensory *proteins* located in the cell wall (*membrane*) catch some of the taste molecules that have been trapped in the mucous coating of the tongue. These special sensory proteins are the *receptors*. A receptor is like an antenna that can identify and pick up a chemical signal, the taste molecule, and send information about this event via the protein through the cell membrane into the cell. Here the signal is intensified by the release of a large number of signal molecules. This process of transmission thus acts as an amplifier. The taste sensory cells are connected to the central nervous system by the nerve cells which then tell the brain that a particular taste has been recorded.

An individual taste sensory cell has receptors for all types of taste. A given taste is, therefore, initially a combination of many individual impressions. It is the brain that sorts it all out and makes a final determination to tell us which taste we have experienced and where in the mouth we have actually sensed it. Empirically we know that we tend to taste sweet things at the front of the tongue, salty things further back, sour ones even further back, and bitter ones at the very back of the tongue.

The question arises as to why we should be able to taste anything. It turns out that we have an abundance of different receptors which are sensitive to bitter taste sensations. It is plausible that they are

学術 **How does water taste?** It is likely that most people would say that it has no taste or, possibly, that water tastes of the minerals which are found in it. But what about completely pure water which has been distilled and demineralised? It tastes sweet. In English, fresh water, as opposed to salt water, is sometimes referred to as 'sweet water'.

Recent research has proven that the sweet taste of pure water is a subjective taste experience, which is due to the fact that the *receptors* in the taste buds which register sweetness are cleansed by the water. This frees up substances bound to the receptors and thereby activates them. The sweet taste is really a psychophysical aftertaste. We know of the same phenomenon from the sense of sight. If we stare at a coloured pattern and then move our gaze to a white paper, we can see imaginary colours on the white sheet.

THE BIOCHEMISTRY OF TASTE

The biochemistry that underlies taste experiences is complicated. Its principal features are as follows. Sweet, bitter, and *umami* tastes are recorded by special protein molecules, the *receptors*, which are located in the *cell membrane* of a taste sensory cell. The receptors can recognize certain types of chemical combinations. When a receptor has identified and latched onto a taste molecule, a signal is sent via the protein that another specific protein (a so-called G-protein) must be bound to it on the other side of the membrane. This sets in motion a cascade of processes that cause certain sodium channels in the cell membrane to open. In turn, this leads to a change in the membrane potential, which can be registered electrically by the nerve cell.

Sour and salty tastes are recorded electrochemically by the receptors of the taste sensory cells, in that hydrogen ions (H^+) and sodium ions (Na^+), respectively, change the membrane potential of the cells. Potassium ions (K^+) can also activate a salty taste sensation. The receptors for sour and salty tastes are transmembrane ion channels.

The biochemical recording of a taste impression is not uniquely chemically determined, in that substances which are chemically widely different can be registered by the taste sensory cells as having the same taste. This means, for example, that not all *saccharides* taste sweet and, conversely, that substances other than sugar can also taste sweet.

As a consequence, the sense of taste is much less refined than the sense of smell, which utilizes different and very specific molecular receptors for each smell substance that can be distinguished.

At present we know for sure that there are several different receptors for bitter tastes and more than one for the sweet. It is assumed that there are also a variety of receptors for each of sour, salty, and *umami* tastes.

there to help us to identify the often poisonous bitter substances which plants use as a protective mechanism.

Sweetness helps us to identify foods that are rich in calories, and salty and sour tastes assist us in regulating the balance of salts and acids. So, where does *umami* fit into the picture? It is not inconceivable that the *umami* taste is an important indicator of food that is rich in proteins.

SMELL

Sometimes one can judge the quality of a good French fish restaurant by the aroma which greets one when one steps inside the door. This informal test does not work when one goes into a sushi bar. Here it is not supposed to smell of anything other than vinegared rice! Some sushi restaurants do, however, also serve cooked and fried dishes, such as battered deep-fried fish, shellfish, and vegetables (*tempura*), grilled skewers of meat and vegetables, as well as a variety of soups and cooked vegetables. There is no doubt that the smell of warm *miso* soup is unmistakeable.

As a sushi meal gives off no aromas, certainly not any whiffs of fish, a guest at such a meal cannot guess by the smell what is going to be served as the next course.

The sense of smell is much more refined than taste perception. Olfactory impressions are airborne, and humans have a reasonably good and acute sense of smell, even if it is much less well developed than that of many animals. Olfactory substances in foods are molecules that have torn themselves free from the food or drink and drift around in the air either on their own or in small droplets of water or fat. An exceptionally large number of molecules are let loose when the food is placed in the mouth. When we act both mechanically, by chewing, and chemically, by adding saliva, on the food, these molecules can break free and are either dissolved in the saliva or move out into the air passages to work their way up to the nose.

Olfaction arises when these molecules come upward in the nose and reach the roof of the nasal cavity where a grouping of specialized neurons is found under a mucus membrane which is covered with small hairlike protrusions or cilia. The process is similar to that involved

学術 WHERE ARE TASTE AND SMELL centres located in the brain? Nerve cells from the olfaction sensitive sensory cells in the nose lead to the olfactory nerve, one of the twelve cranial nerves, which forms a bulbous protrusion at the very front of the underside of the brain (*piriform cortex*). Taste is sensed in two areas in the middle of the front part of the brain (*post-central gyrus*).

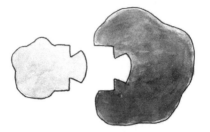

Each type of olfactory molecule fits, like a key in a lock, into its special receptor in the nose.

学術 THE SENSORY CELLS IN THE NOSE that detect smells have about 2000 different receptors, which can each detect one, and only one, type of olfactory molecule.

学術 TEXTURE MATTERS. The Japanese put much more emphasis on the texture of their food than people in the Western world and they have a long list of phrases that characterize various aspects of the texture of foodstuff. In fact, they speak not only about mouthfeel (*kuchi atari*), but also of tonguefeel (*shitazawari*) and tooth resistance (*hagotae*). Crustaceans and seaweed are often eaten because of their special texture.

in taste perception. A smell molecule is recorded by binding itself to a *receptor*, which is a protein located in the mucosa of the sensory cell. When the receptor recognizes a specific chemical substance, a large number of signal molecules on the inside of the membrane are released and, after a series of intermediary steps, the amplified signal reaches that part of the brain where the olfactory centre is found.

There are several thousand different olfactory receptors. But one can detect and differentiate many more disparate smell sensations because an odour can be a composite of signals from a number of individual receptors.

The sense of smell is much more sensitive than the sense of taste. We all know the feeling that food tastes different when we have a cold and our nose is blocked. From a purely chemical perspective, there is no quantitative change in the taste, and there is no physiological connection between taste and smell perception. What comes into play is that the sense of smell of a person with a cold is diminished because the nasal mucosa is temporarily thicker and the combined taste and smell impression is consequently altered.

It is characteristic of sushi and the dishes described in this book that, with only a few exceptions, they are eaten cold, that is to say, normally at room temperature. As a consequence, the substances that give rise to olfaction are less prominent than they are in dishes that are served warm. This is an important factor which has a major influence on the presentation of sushi and related dishes. They depend on their visual impact to make the mouth water.

MOUTHFEEL

The properties of food in the form of texture, consistency, body, and temperature elicit physical impressions which can partially be characterized both objectively and quantitatively. These various impressions have been synthesized into the modern concept of mouthfeel, an expression commonly used in the testing and evaluation of food and drink. The perceptions that contribute to mouthfeel are overwhelmingly physical and mechanical in origin, but one cannot divorce them completely from the sense of taste.

Simple characterizations of mouthfeel are often carried out using a qualitative scale of opposites, such as soft-hard, tender-tough, dry-creamy, floury-waxy, and crumbly-elastic. The last-named, elasticity, is also related to *viscosity*, an indication of the degree of fluidity of a substance.

Some taste sensations can also be linked to a mechanical feeling during the chewing process, which is often accompanied by sounds that are heard either directly or indirectly via the jawbone and cranium. Examples are crackly, crispy, squeaky, and crunchy noises.

In the preparation of *nigiri*-zushi, that is to say, balls of cooked rice topped with fish or shellfish, one should try very hard to ensure that both the rice and the fish have the same degree of softness. Otherwise, either the piece cannot be bitten in half (which some purists decry as a sacrilege), or else the rice and fish separate from each other on the way to the mouth or in the mouth.

It is a cardinal principle of Japanese cuisine that much of it, including sushi, seaweed, and *sashimi* (bite-sized pieces of sliced raw fish or shellfish), is to a great extent consumed simply for its texture.

A taste sensation can occasionally be enhanced by a chemical irritation of the tongue and the mouth. We know the effect produced by mustard, chili peppers, and garlic. The Japanese horseradish, *wasabi*, which is used in sushi dishes, produces a similar stimulation, which can impart an agreeable impression at the same time as it brings out the taste of another ingredient in the food, in this case, the fish.

The feeling of irritation in the mouth and on the tongue, which is called forth by substances such as the ones above, is due to the fact that these plants contain substances that damage the cells. The damage registers in the brain as pain.

Another chemically dependent mouthfeel is *astringency*, with which we are familiar from foods with an abundance of *tannin*, such as tea, unripe bananas, or immature red wine. The tannin binds to the *proteins* in saliva, which normally allows food to slide around easily on the tongue and in the oral cavity. The result is a taste sensation that is biting, dry, and chafing. Irritation and astringency can be perceived as either pleasant or unpleasant, depending on the circumstances.

"SOMETHING FROM THE SEA & SOMETHING FROM THE MOUNTAINS"

What fish feel
What birds feel, I do not know
The year is drawing to a close
Matsuo Bashō (1644-1694)

'THE FRUIT OF THE SEA': FISH AND SHELLFISH

The oceans of the world are filled with a vast abundance of fish and shellfish which are very suitable for the preparation of sushi. Some can be eaten raw, while others must first be cooked, marinated, or fermented. There is a wealth of colours – white fish, red fish, shiny fish – and a variety of patterns and textures. As a bonus, fish and shellfish are healthy foodstuffs, low in calories, and rich in proteins and superunsaturated fats.

FISH, SHELLFISH, AND ECHINODERMS

The genus fish encompasses a large group of vertebrates which live in water and respire through gills. It includes bony fish, cartilaginous fish, and round mouthed fish. Only bony fish, of which there are more than 25,000 different living species, are used for making sushi. Depending on the fish, the muscles (fillet), skin, and/or roe may be used as raw ingredients.

In zoological terms, shellfish is a generic term for invertebrates with external shells. The genus includes crustaceans and molluscs with a shell, such as *bivalves* and *gastropods*. Gastronomically speaking, shellfish is a catch-all designation for both of these categories even though, in contrast to the strictly scientific classification, it also encompasses those molluscs which do not have shells.

Molluscs comprise a large phylum of invertebrates which includes at least 100,000 living species. Most have external shells, like clams, oysters, and escargot, or are *cephalopods*, such as squids and octopuses, which have either a reduced internal or external shell, or no shell at all. Molluscs are well suited for preparation as sushi.

Crustaceans constitute a very extensive class of invertebrates with about 54,000 different living species. Their heads are usually fused with their bodies, as is the case for shrimp, and many have clearly segmented bodies, river shrimp being a good example. The crustaceans used for sushi are *decapods*, that is to say, animals with ten legs.

Echinoderms are a phylum of invertebrates, of which there are about 6,000 different surviving species, including the classes sea urchins, starfish, and sea cucumbers. Their distinguishing characteristics are a fivefold radial symmetry and an exoskeleton. The only ingredients derived from the echinoderm phylum used in sushi cuisine are sea urchin reproductive organs, commonly referred to as roe.

Why are fish muscles soft?

Most people have probably wondered why fish muscles are so soft. One can easily poke a finger through a fillet of a bony fish like mackerel, salmon, or tuna, whereas it is not possible to do so on a piece of meat from a terrestrial animal, be it beef, pork, or poultry, as it is too tough. How can it be that the muscle of a fast-swimming predator like a tuna is soft while the flesh of a slow-moving ruminant like an ox is chewy and firm?

The explanation is that fish, in contrast to terrestrial animals, do not need to use their muscles to bear their own weight or to maintain their body shape. Most fish have the same density as the water in which they live, so their effective weight is negligible. As fish need to use muscle power only to move around, they simply have to work much less than terrestrial animals.

Salmon muscle with short red fibres held together by pale stripes of connective tissue.

学術 **When a fish becomes rigid.** The firmness of the muscles of a dead fish is very dependent on the length of time that elapses until *rigor mortis* sets in, typically about six hours after it has died. *Rigor mortis* releases the calcium ions of the *proteins* embedded in the muscle fibres and the fibres then are locked together in a sort of contraction.

Rigor mortis can be delayed for a period of up to a few days, by deep-freezing the fish immediately after it is caught. If the fish has become tired out or stressed under capture, *rigor mortis* might set in earlier.

Once the process of *rigor mortis* has run its course, the *enzymatic* decomposition of the fish commences, the muscle fibres separate, and the connective tissue is loosened. It is precisely at this point that it is ideal to consume the fish, among other ends to make *nojime* sushi, the type that is made from fish which is not kept alive after it is caught. The opposite is *ikijime* sushi, prepared from fish with firmer muscles because it is kept alive until the last moment and used before *rigor mortis* can set in.

On the other hand, fish have to be able to mobilize their muscle power very quickly, for example, to evade a predator or to capture their own prey. Furthermore, their strength has to be sufficient to propel them through a liquid medium, water, which is much denser and provides more resistance than land animals encounter from air. For these reasons fish muscles are constructed differently from those of terrestrial animals, a difference that results in their being very tender.

MUSCLES

Muscles consist of bundles of muscle fibre, connective tissue, and fat. The fibres are composed of two types of protein: *myosin* and *actin*. Myosin is a *molecular motor*, which can slide over the actin and cause muscle contraction. Myosin needs energy in the form of *ATP* (*adenosine triphosphate*) in order to function. The connective tissue is made up of *collagen*, which is a spiral-shaped protein. In the connective tissue it is wound together into triple helical strands, much like a coiled rope.

The muscle of terrestrial animals is firm and tough; this is explained by a number of factors. For example, in mammals muscle fibres are very long and extend through the entire muscle. In addition, the collagen is relatively stiff and has to reach a temperature of about 60-70°C before it starts to melt to a *gelatinous* state. Finally, the muscle fibres are bound quite tightly to the skeleton by connective tissue, of which there is much more than there is in fish.

In bony fish the muscle fibres are short and held together by a looser form of *collagen* which *denatures* and turns to *gelatine* at a mere 40°C. Also, the adhesion of the collagen to the bones is weaker. As a consequence, fish muscles are soft and, when the collagen is heated a little, they flake and come off the bones very easily.

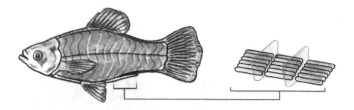

The picture illustrates the structure of the striated muscle tissue in a bony fish. Individual muscle fibres are less than a tenth of a millimetre in diameter. In contrast to the very long fibres of which the muscles of terrestrial animals are composed, the fibres in fish muscles are divided into short layers, *myotomes*, which are typically from a few millimetres to one centimetre long. When a cooked fish flakes apart, one can easily see this stratification. The myotomes are held together by fragile layers of connective tissue: *myosepta* along the fibre bundles and *myocommata* across them. Myocommata extend from the innermost layer of the skin (*dermis*) to the bone. They are arranged in a sort of zig-zag structure which gives fish muscles their special texture. This construction is easiest to identify in salmon on account of its distinctive red colour, caused by a pigment that accumulates in the fat deposits in the muscle fibres. The whitish connective tissue stands out against it in clear contrast.

Slow and fast muscles

Muscles that have to work all the time in terrestrial animals and to be in constant motion in fish are classified as slow muscles. They need a continuous oxygen supply to produce energy in the cell *mitochondria* and to burn fat. A *protein* called *myoglobin* is responsible for the transport of oxygen within the muscle tissue. Myoglobin is a relative of *hemoglobin,* which carries oxygen from the lungs, via the bloodstream, to the muscles, where the myoglobin takes over. The oxygen is expended in the conversion of *glucose* to form the energy (*ATP*) that is required to make muscles contract. Glucose, which also circulates in the bloodstream, is produced from the reduction in the liver of *glycogen*, a large branched *polysaccharide* made up of glucose units.

Muscles that need to work only in short bursts and can yield a great deal of strength over short periods of time are classified as fast muscles. As they often are not able to wait for the arrival of oxygen, they contain glycogen, which is available for immediate use in their own energy depots. The reduction of glycogen can take place in the absence of oxygen and results in its being converted either to *lactic acid* or directly to carbon dioxide and water. For this reason the fast muscles can function only for a limited time and in the aftermath must receive oxygen in order to break down the lactic acid.

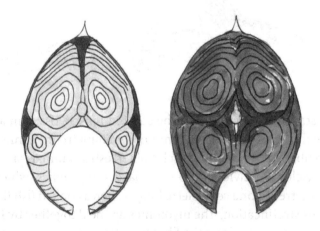

The figure above shows a cross-section through the muscle tissue of fish, which is predominantly made up of either fast muscles (light) or slow muscles (dark and red). The reason that sharks have many slow muscles and dark flesh is that their specific gravity is greater than that of the water which surrounds them. Hence they have to be in constant motion in order not to sink to the bottom of the ocean.

WHY ARE SOME FISH MUSCLES WHITE?

It is a common, but erroneous, belief that the red colour of meat is due to the presence of blood in the muscles. The protein *hemoglobin*, which transports oxygen from the lungs to the muscles, is red, but it is not found in the muscles.

Since the muscles' own proteins are nearly without colour, it is another substance that makes some muscle types dark and red. This substance is the red protein *myoglobin*, which in the so-called slow muscles is responsible for transporting oxygen within the muscle tissue. The slow muscles take care of work that has to be carried out on an on-going basis, namely, continuous swimming, and this is why they are dark and red.

In contrast, the so-called fast muscles undertake tasks of short duration, that is to say, smaller but rapid and demanding movements and, on occasion, the slapping of fins and tail. These muscles do not contain myoglobin and instead use the colourless *polysaccharide glycogen* to produce energy. This is why the fast muscles are colourless or whitish.

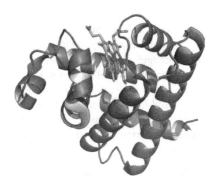

学術 **MYOGLOBIN** is the *protein* that carries oxygen in the slow muscles. The protein contains an iron atom that can bind oxygen and in this way transport it around in the muscle. When it is not bound to oxygen, *myoglobin* is purple.

Each myoglobin molecule can bind one oxygen molecule, becoming oxymyoglobin, which is bright red. In the illustration above, the part of the molecule that is indicated in blue is called a heme-group, which consists of a porphyrin ring that surrounds the iron atom. Because of the oxygen atom bound to the myoglobin, the iron atom after a period of time changes its oxidation level (from Fe^{++} to Fe^{+++}), thereby producing saturated myoglobin, which has a brownish colour. This colour transformation can be reversed by changing the amount of oxygen present.

When myoglobin is warmed to a temperature of over 40°C or when it is exposed to acid, the protein again turns a brownish colour, but this time the process is irreversible. This is because the protein is *denatured* and unwinds, setting free the heme-group. In the kitchen this can be illustrated by two examples: a beautiful red myoglobin filled tuna fillet turns a drab grey-brown when steamed slightly, whereas a white turbot fillet, which contains only a little myoglobin, keeps its white colour when it is cooked.

Fillets of fish with red slow muscles (tuna) and white fast muscles (halibut).

学術 **Why are the muscles** of wild salmon and sea trout pink, red, or orange? The colour is due to the accumulation of the natural pigment, *astaxanthin*, in their fat depots. It is found in plankton, which is eaten by the tiny crustaceans, which in turn form an important part of the food chain of these fish.

In the intact shells of the crustaceans, the astaxanthin is bound to a protein (*crustacyanin*). In this form, its colour is more blue-green or reddish-brown than red, as we know from seeing live shrimp, lobsters, crayfish, and crabs. The crustacyanin is *denatured* when the fish digests the shells. This frees the astaxanthin, allowing the pigment's own reddish orange colour to take prominence. The same denaturing process occurs when crustaceans are heated, a phenomenon familiar to all who have cooked them.

Salmon and sea trout which are farmed, and as a result do not have access to wild crustaceans, normally have paler muscles. The aquaculturist can induce a redder colour by feeding the fish crustacean shells or by adding a chemically related food colouring (*canthaxanthin*) to their feed.

Normally bony fish have both slow and fast muscles and one can see by the colours where they are in the body of the fish. The muscles in the tail and along the fins, which are more often in motion and which have to provide power on a sustained basis, are typically darker than the others. Predatory fish constantly in search of prey will have a greater proportion of slow muscles and, consequently, more red muscles. Tuna is an example of a fish with very red muscles. Flatfish, which lie practically still on the sea-bed, have a preponderance of fast muscles, which are only activated in short bursts when a victim passes by. So these fish have transparent or white muscles.

Some fish with slow and, hence, dark and red muscle fibres, for example, tuna, have a large fat content in some parts of the body, causing those muscles to be pale or pink with whitish fat marbling. The light, fat marbled tuna belly (*toro*) is sought-after for sushi on account of its pleasant, soft consistency and distinctive taste.

Salmon has a red pigment in its muscles

Salmon and sea trout have pink, red, or orange muscles. Their colour is not due to the presence of *myoglobin*, but rather to a pigment called *astaxanthin*. Astaxanthin is a *carotenoid* and is chemically related to the pigment that gives carrots their characteristic orange appearance.

The protein complex *crustacyanin*, which is bound to *astaxanthin* molecules, is carrot coloured in the illustration. When the complex breaks apart, the astaxanthin molecules are freed and the colour changes from blue-green to reddish orange.

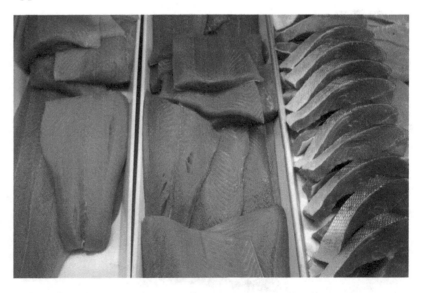

These salmon fillets are red from the accumulation of *astaxanthin* in the muscles. The pigment comes from the tiny crustaceans which are the mainstay of their diet. Without astaxanthin a salmon fillet would be white, because it consists primarily of fast muscles.

FISH DO NOT SMELL FISHY!

The classic, disagreeable 'fish odour' is not the smell of the fresh fish. It is due to certain chemical changes in the dead fish, changes that are set in motion by its own *enzymes,* as well as by attacking microorganisms. Fresh fish and shellfish smell just like a pleasant sea breeze – subtle, cool, and with a whiff of salt water and iodine.

Very fresh fish is also said to have a smell resembling that released by the leaves of plants when they are crushed. Both plants and fish contain large quantities of polyunsaturated *fats* (*linoleic acid* and *alpha-linolenic acid*), which are oxidized by special enzymes (*lipoxygenases*). This reaction creates a transient aromatic olfactory substance which we identify with the smell of plants. In addition, salt water fish accumulate *bromophenols,* which are found in the marine algae eaten by the fish and by the smaller marine animals in their food chain. As there is no parallel source of bromophenols in fresh water, river and lake fish do not give off that same agreeable, refreshing ocean tang which these substances impart.

A dead fish is very quickly attacked by its own *enzymes,* especially digestive enzymes. What is special about them is that they are designed to function at relatively low temperatures. Fish have the same body temperature as the water in which they live, in contrast to such animals as mammals whose body temperature is higher than that of fish and is tightly regulated. For example, the enzymes in the digestive system of an ox function optimally at around 37°C, and their activity decreases very markedly when they are cooled or frozen.

On the other hand, the enzymes in fish function very well at the temperature levels to which our refrigerators are set and even in frozen fish they continue to be somewhat active. For this reason, fresh fish should preferably be stored on ice. The digestive enzymes of fish are very aggressive and designed to break down other organisms at low temperatures. After all, the stomach and intestines of a fish must be able to deal with the shells, bones, and scales of the small fish and crustaceans which the fish digests after having swallowed them whole and unchewed. This is why it is so important to remove the pharynx and innards of a fish as soon as it is caught.

学術 LOOK YOUR FISHMONGER straight in the eye. It can be difficult to figure out for yourself how fresh a fish is, especially if it is already filleted. My advice is: look your fishmonger straight in the eye and ask him or her point blank. Buying fish and shellfish for sushi and *sashimi* is a matter of trust.

学術 LOOK YOUR FISH in the eye, too. Fresh fish have clear, black, and resilient eyes, with a shiny surface. A gentle poke on the side of a fresh fish should not leave a depression. But the colour of the gills is the best indication of freshness. The gills are the respiratory organs of the fish and *hemoglobin* in the blood colours them bright red. They are the first part of the fish to be subjected to *enzymatic* breakdown and over-oxidization; consequently, they spoil much more quickly than the muscles. Just as with *myoglobin,* over-oxidization causes the hemoglobin to lose its red colour and turn brown.

学術 **Mackerel**, like its relative the tuna, is an active hunter in constant motion. This is why it has many slow muscles and a large quantity of active *enzymes*, which contribute to its rich taste, but also cause the fish to decompose very quickly if it is not cooled or frozen immediately after it is caught. Even in the frozen state, its enzymes are partially active – consequently, a fat mackerel does not last nearly as long in the freezer as does a lean plaice.

In the fish store, mackerel and herring are left ungutted, as the innards should be removed only when the fish is about to be prepared. The reason for this is that it is virtually impossible to gut the fish without leaving some of the digestive enzymes behind on the cleaned fish. They are so active that even in small quantities they quickly cause decomposition.

学術 **Fish is easy to digest.** Active *enzymes* in fish, especially in raw fish, are a contributing factor in making fish more digestible than meat from terrestrial animals.

学術 **Spots on a fish** are an indication that it was handled too roughly either in the process of being caught or under transport. Some flatfish, however, have naturally prominent spots, even on their white undersides. But being hit or squeezed can cause dark red or brown spots, which are a sign of blood seepage and breakdown of *myoglobin*. Apart from being an aesthetic disadvantage, such spots will accelerate the *enzymatic* decomposition of the fish and, in the process, affect its taste. This is why fish should be handled as little as possible, and with great care, from catch to consumption.

The catabolic by-products resulting from oxidation of the fish's polyunsaturated *fatty acids* are also agents of decay and produce a disagreeable smell, especially in the case of oily fish.

Bacteria and other microorganisms living on the skin and in the digestive system will break down a dead fish, in the process forming *trimethylamine*. Trimethylamine is a basic chemical compound with an ammonia-like odour that we normally associate with spoiled or rotten fish. Compounds containing sulphur may also be formed, contributing other unpleasant odours. The onslaught of the bacteria starts only after *rigor mortis* has run its course, typically about six hours after the fish has died. *Rigor mortis* can be delayed for a few days by freezing the fish as soon as it is caught.

Cross-sections of fillets of oily fish, such as tuna, can display a rainbow-like effect due to the migration of fats contained in them to form a very thin oily film on the surface. This indicates that the fillet has been lying around for about a day, but is not necessarily a sign that the fish is not fresh enough to eat.

A coating of slime covers all types of fish to a greater or lesser degree. An odourless, clear, and even layer of slime is a sign of freshness. If the layer of slime smells or is lumpy and damaged, it might indicate that the fish is not fresh or that it has been handled too roughly under transport.

Fish odours can, to a certain extent, be eliminated from reasonably fresh fish by rinsing the surface of the fish with clean water, to which a little lemon juice or a bit of wine vinegar may be added.

How do fish taste?

As a general rule, the taste of a fish is closely related to its fat content. More *fat* equals more taste, but it is not a given that this will be a pleasant taste. A high concentration of fat can result in a strong taste of fish oil. Fish from cold and temperate waters are often more flavourful than fish from warm and tropical seas. The reason for this is that colder, and generally more turbulent, waters have a richer and more varied stock of plankton and a greater concentration of small suspended mineral particles (*colloids*), both of which comprise the diet of the small denizens of the sea on which the larger fish prey.

The fat content of a fish is also contingent on which part of the water it inhabits. Fish that live in deep oceans, as do flounders, typically have a low fat content. On the other hand, predators like tuna and mackerel that operate near the surface of the water are very oily.

Ocean fish and shellfish live in water in which the salt concentration can be as high as 3%. In order to maintain the correct *osmotic* balance so that the cell fluid does not leak out, these animals have to concentrate substances that will retain water within their cells. These substances are certain free *amino acids*, for example, *glycine* and *glutamic acid*, which have a somewhat sweet taste. It is glutamic acid, in the form of *monosodium glutamate* (MSG), that brings out *umami* taste. Mackerel is an example of a fish which has large quantities of MSG together with other palatable amino acids.

The greater the salinity of the water inhabited by a fish, the more its cells have to store up substances to counteract the osmotic effect. Consequently, fish from very salty oceans have a distinct sweet-spicy taste.

Completely fresh fish also contain *nucleotides,* which help to impart *umami* taste, especially the nucleotides *inosine monophosphate* (IMP) and *guanosine monophosphate* (GMP) that are created when the cells of the fish have to produce energy by breaking down ATP (*adenosine triphosphate*). These savoury substances disappear gradually after the fish has died.

Fish which live in fresh or brackish water have less need to balance the osmotic effect by concentrating amino acids in their cells and, therefore, they have a milder taste than ocean fish. River fish normally have a stronger taste than lake fish because they have to move more vigorously to navigate through current-filled waters, resulting in a better developed musculature.

学術 **A HAPPY FISH TASTES BETTER** than a stressed fish. The fast musculature of fish contains *glycogen*, which can quickly be broken down to supply *glucose* to provide energy in case the fish has suddenly to splash its tail or quickly swim away to evade an attacker. The glycogen supply is sufficient only for muscle functions of short duration, so the fast muscles rapidly become tired when the glycogen is almost depleted. Unused glycogen in a dead fish is transformed into *lactic acid*, which has a preservative effect. A stressed fish that has exhausted its glycogen by trashing around in a fish net or by fighting to free itself from a fish hook on the end of a line will, therefore, neither keep as well nor taste as good as a happy, unstressed fish.

In some fish farms the fish are transferred to very cold water, which is just above the freezing point, before they are slaughtered. At such low themperatures, the fish become very inactive and they do not become stressed before they are killed.

Osmosis

Osmosis is a physical phenomenon which occurs across a barrier, for example, a *cell membrane*, that is permeable to water but impermeable to other larger molecules, such as salt or sugar. The imbalance that arises therefrom is compensated for by seepage of some of the water to the side on which the large molecules are found. The extent to which this happens is proportional to the degree to which these molecules have an affinity for water. The osmotic effect is countered by osmotic pressure, which arises across the membrane. Osmosis is vital to plants' ability to absorb water from the ground into the root system and up through its stems.

If the osmotic pressure is great, the membrane can actually burst. In the case of a cell membrane, the cell will then die. For example, this takes place when fish or vegetables are cured in brine, causing both their cell walls and those of microorganisms associated with them, if any, to burst.

In order to minimize osmotic pressure, cells store their nutrition and fuel in the form of *polymers*, that is to say, partly as proteins rather than as *amino acids* and partly as *polysaccharides* instead of *glucose*. This is possible because the osmotic pressure is not dependent on the size of the molecules, but solely on how many of them there are.

The unpleasant smell of fish that is not fresh is attributed to the chemical substance *trimethylamine*, which is formed by bacterial breakdown of *trimethylamine oxide* in the dead fish. Trimethylamine oxide, which in itself is odourless, is utilized by the cells of the fish to balance out the osmotic pressure due to the saltiness of the ocean water. Fresh water fish have only a little trimethylamine oxide in their cells, 0.5 mg/kg compared to 40-120 mg/kg in salt water fish.

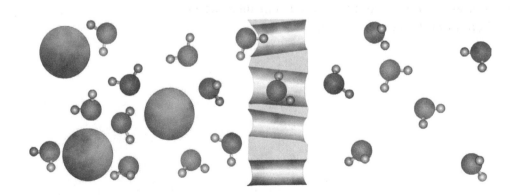

Fish and shellfish. Woodblock print by Utagawa (Ando) Hiroshige (1797-1858).

学
術
WHICH FISH ARE CAUGHT the most on a global basis? Salmon and trout are the leading fresh water fish, the herring family (herring, sardines, and anchovies) make up the largest proportion of the salt water catch; shrimp are the most harvested crustaceans; and octopuses are most prominent among the molluscs. About 22% of all fish and shellfish are sold fresh, 24% are frozen first, and 24% are preserved (salted, smoked, marinated, or canned). The balance, ca. 30%, is made into fish oil or fish meal. (Belitz et al., *Food Chemistry*, 2004)

FISH AND SHELLFISH ARE NUTRITIOUS FOOD

Fish and shellfish are a rich source of *proteins*, *vitamin* B, and a series of minerals, such as calcium and iodine. In addition, oily fish and fish livers have significant amounts of vitamins A and D, which are fat soluble, as well as of vitamins E and K. Fish, especially oily fish, also contain the very important polyunsaturated fats, the *omega-3 fatty acids*. These have low melting points, allowing them to remain in the liquid state even at the low temperatures found in some oceans.

Two types of omega-3 fatty acids are vital nutrients for humans: *eicosapentaenoic acid* (EPA) and *docosahexaenoic acid* (DHA), both superunsaturated fatty acids with five and six double bonds, respectively.

DHA and EPA play a major role in building up our central nervous system, especially the brain and its visual forefront, the retina of the eye. In contrast to the saturated fats ingested from animal products, DHA and EPA prevent cardiovascular disease and cancer and help to regulate the *cholesterol* count in the bloodstream.

Our bodies cannot create DHA and EPA on their own and, therefore, fish and shellfish are a major source of these fats. For example, DHA makes up 50% of the fat content in the muscle tissue of a salmon, whereas it makes up only 0.2% in beef. Fish ingest omega-3 fats in their diet, which either directly or indirectly consists of phytoplankton found in the oceans. For the same reason, farmed fish have a lower proportion of omega-3 fats and fresh water fish even less.

Fish oil capsules from oily cold water fish.

Check the epa and dha content of your fish oil

The *omega-3 fatty acids, eicosapentaenoic acid* (epa) and *docosapentaenoic acid* (dha), are the most important components in the fish oil you can buy as a liquid or in gel capsules. Commercially, it is normally produced by pressing the oil deposits out of fish muscles. The best readily available oils contain ca. 32% epa and ca. 22% dha by volume.

epa and dha are superunsaturated *fatty acids,* which contribute greatly to our well-being, and are essential elements of our nervous system, brain, and retina. They help to prevent cardiovascular disease and can lower the *cholesterol* count in the bloodstream.

In addition to dha and epa, fish oil contains a series of other, undesirable oils. An *antioxidant*, for example, vitamin E, is routinely added to prevent the unsaturated fatty acids from *oxidizing* (becoming rancid) and thereby losing their beneficial properties.

One can avoid the undesirable oils by buying a carefully purified fish oil that has a greater concentration of, and cleaner, dha and epa. A new product of the biotechnology industry has recently appeared on the market. It is not derived from fish at all, but is enriched with dha extracted from algae, which are the original sources of unsaturated fatty acids for the fish. The fish have been sidelined as an intermediary.

It is recommended that one should ingest the necessary amounts of superunsaturated fats from eating fish, as the extracted oil does not contain as much dha and epa as the actual fish muscles. Furthermore, the superunsaturated fats in fish muscles are predominantly in the form of *phospholipids*, whereas in fish oils they are mostly in the form of *triglycerides*. The significance of this is that lean fish like cod can contain just as great a quantity of dha and epa in the form of phospholipids as do oily fish like salmon. Unfortunately, information available regarding the phospholipid content of the different types of fish is still quite limited.

The recommended daily intake of dha and epa is equivalent to 150 to 200 grams of oily fish. Few people manage to consume this much, so fish oil is a good alternative. I myself take 3 grams of fish oil on those days when I am not eating fish. Preferably, the fish oil should be taken together with something else that contains fat, for example, milk or yogurt, in order to *emulsify* the fish oil in the stomach and intestines. This emulsification process is very important as it gives the superunsaturated fatty acids a chance to be transported across the mucous membrane of the intestines and enter the bloodstream.

学術 **WHERE IS THE FAT IN A FISH?** The fat is located in special fat cells just under the skin, as well as in the connective tissue in between the muscle *myotomes*. In addition, fat is found right among the microfibres in the muscle bundles. The muscles of the abdomen contain more fat than those on the side. An example of this is tuna belly (*toro*), which is regarded as the greatest sushi delicacy.

学術 **THE CALORIE CONTENT OF FISH** is relatively low. Lean fish generally have 100 calories per 100 grams and the oiliest ones, such as mackerel and tuna belly, have about twice as many.

学術 **TAURINE** is an *amino acid* that does not become a component of *proteins*, but is nevertheless very important, especially for the formation of gall salts, which bind *cholesterol*. Investigation of its role seems to indicate that it helps both to isolate cholesterol and to lower the cholesterol count in the bloodstream.

Taurine is found as a free *amino acid* in a number of fish and shellfish. It is especially abundant in red tuna, cephalopods, shrimp, and some bivalves, for example, scallops.

OILY AND LEAN FISH

There is extraordinary variation in the fat content of different types of fish, from 0.5% in cod to 30% in a plump herring. Also, the fat content varies with the seasons and is dependent on the spawning cycle of the fish.

The *cholesterol* content of bony fish is low, typically 0.05%. Bivalves have a correspondingly low, or even more negligible, cholesterol count. Cephalopods, especially cuttlefish, and crustaceans contain a fair amount of cholesterol. There are also significant quantities of cholesterol in roe, for example, that from salmon (*ikura*) and sea urchins (*uni*).

The relatively high cholesterol content in cephalopods can, to a certain extent, be offset by a correspondingly high level of the *amino acid taurine*, which binds the cholesterol.

THE BEST PART OF THE FISH

The various parts of a fish have different fat contents. The fattiest, and, hence, the softest, part is near the head and stomach, at the front of the fish. Conversely, the tail is the leanest and has the strongest taste because that is where the most active muscles are found. Of course, that also makes it the toughest part. The best compromise in terms of tenderness, fat content, and taste is to choose the middle of the fish, generally the most desirable section for making sushi and *sashimi*.

FISH WITH BONES

The structure and size of the skeleton and bones in bony fish are dependent on where they live. Because the specific gravity of salt water is greater than that of fresh and brackish water, a fish is more buoyant in the former. This is why salt water fish can get away with having a heavier skeleton and bigger bones than fish from fresh or brackish water, which have many small, thin bones.

Fish bones are very suitable for making soup stock. In fish with a lot of connective tissue, like mackerel, the bones can be deep fried and turned into a crunchy snack.

FAT CONTENT OF FISH AND SHELLFISH

FAT CONTENT OF FISH

Lean fish (0.5%-3%)	Moderately oily fish (3%-7%)	Oily fish (more than 7%)
Cod 0.6%	Tilapia 3%	Atlantic salmon 4-15%
Haddock 0.9%	Sea trout 3%	Herring 4-31%
Yellowfin tuna 1%	Turbot 3%	Mackerel 7-34%
Snapper 1.3%	Bluefin tuna 5%	Greenland halibut 16%
Lemon sole 1.4%	Pacific salmon 7%	Eel 19-41%

The numbers in the table indicate the percentage of fat in the edible parts of the fish. As indicated, the fat content of some species of fish can vary widely. (Sources: McGee (2004), p. 184; Belitz et al. (2004), Table 13.5, p. 627; USDA National Nutrient Database; BC Seafood OnLine; www.aboutseafood.com; the Danish Food Composition Databank [www.foodcomp.dk])

SUPERUNSATURATED FATS IN FISH

Fish species	DHA	EPA
Sockeye salmon	1.30%	1.70%
Mackerel	0.65%	1.10%
Tuna	0.63%	1.70%
Atlantic salmon	0.18%	0.61%
Plaice	0.09%	0.09%
Cod	0.08%	0.15%

Percentage by weight in fish muscles of the two superunsaturated *omega-3 fatty acids: eicosapentaenoic acid (EPA)* and *docosapentaenoic acid (DHA)*. (Source: Belitz et al. (2004), Table 13.8, p. 630)

FAT CONTENT IN OTHER SEAFOODS

	Total fat %	Cholesterol %
Salmon roe (*ikura*)	14.0%	0.50%
Sea urchin roe (*uni*)	3.0%	0.50%
Shrimp (*ebi*)	1.7%	0.15%
Cuttlefish (*ika*)	1.4%	0.20%
Octopus (*tako*)	1.0%	0.05%
Scallop (*hotategai*)	0.7%	0.04%

(Sources: C. D. Bledsoe & B. Rasco, Caviars and fish roe products. *Crit. Rev. Food Sci. Nutr.* **43**, 317-356 (2003); USDA National Nutrient Database; BC Seafood OnLine; www.aboutseafood.com)

Flying fish (*tobiko*)

Salmon (*ikura*)

Smelt

Coloured lumpfish

Lumpfish

Fish roe

For many people, fish eggs, commonly referred to as roe, are the most sought-after part of the fish. Most fish produce an enormous quantity of eggs, which each contain the germ of a new individual together with a 'box lunch' of nutrients to sustain its initial development. The egg is, therefore, rich in *proteins* and *fats*. The fat content is substantial, typically 10-20%, and the *cholesterol* content is high, about half a percent.

The colour of roe is due to pigments in the fat drops in the egg yolk and changes from one species to another. It can be yellow, red, green, or black. For example, a *carotenoid* (*astaxanthin*) is responsible for the reddish orange colour of salmon roe, while the blackness of sturgeon eggs (caviar) is caused by *melanin*.

A little salt is often added to fish roe to bring out the desirable taste of the *amino acids*.

For sushi, the most highly prized roes are from sea urchin, salmon, flying fish, herring, and sturgeon. Those wishing to prepare it at home might find that fish eggs from lumpfish, smelt, sea trout, and salmon are the most readily available.

Tsukiji – fish on an epic scale
On an excursion to the biggest fish market in the world

A taxi was waiting for us outside the hotel at around 6 in the morning. We were about to visit Tsukiji to experience the biggest fish market in the world. The action starts very early in the day and it had already been open for over an hour when we arrived at 6:15. I was with my Finnish colleague and good friend of many years' standing, Paavo Kinnunen. We were almost at the end of a ten day trip around Japan, where we had scientific discussions with colleagues in universities and organizations, gave talks, and spent a weekend at Arai Ryokan, a traditional Japanese inn located by the hot springs in Shuzenjii. We both have a weakness for Japanese food, especially sushi, and we were determined not to leave Tokyo without having seen the source of the fresh fish we had been eating.

Tsukiji is located in the center of Tokyo, just a short distance from Ginza, the trendiest boutique shopping area in the city. In addition to trading in fish, Tsukiji is a wholesale market for other fresh products – meat, fruits, vegetables, and flowers. As we approached the market, the tempo in the streets picked up. There is constant traffic – everything from small vans to huge freight trucks. As we were arriving a little on the late side, we mostly encountered vehicles already leaving the market, on their way to the many stores and restaurants that were awaiting new deliveries.

The taxi driver let us off at a street corner by the outer market. It is a bit like a small neighbourhood surrounding the market itself. Here you can buy, among other goods, porcelain, ceramics, and a variety of kitchen utensils, such as handmade knives. In addition to all the little stores, trade is also carried out from stands located right on the street. The frenetic pace and the frantic level of activity struck us right away. To the uninitiated it all looks terribly confusing. We had to leap for our lives to avoid being hit by one of the many motorized carrier bicycles and fast trucks which zip around with boxes of fish.

Our destination was the inner market, the actual fish market, an enormous complex of interconnected halls. Inside it the volume of traffic is just as intense, but here the transport is by means of small handbarrows, which are pushed or pulled along the narrow paths in between the booths where the fish are sold. People run in all directions. They have no time to lose because they are dealing with expensive products whose quality and value depend entirely on their freshness.

The seven big wholesalers in the market put their wares on display for viewing very early, at around 3 in the morning, and the auction begins at about 5. The tuna auction starts at 5:30; first the fresh tuna are sold, then the frozen. The fresh ones are all gone in less than 15 minutes, with the rest of the auction essentially wrapped up within the hour. As soon as the auctions are completed, the smaller wholesalers bring their purchases to their stalls in the market, where the fish is cut up, packaged, and resold to retailers. At this point the race is on to get the fresh goods from the market to the shops and restaurants as quickly as possible.

We came to look at fish and were not disappointed. Never have I seen such a quantity and variety of fish. There are fish and all manner of other good things from all the oceans of the world.

Fresh and frozen fish. Shellfish. Octopuses. Fish roe. Mounds of sea urchins (*uni*) made our mouths water. Seemingly endless rows of fish in boxes and large frozen fish lie on the floor, one enormous hall after another. The market is huge. Over 50,000 people work there, and at least 100,000 customers and visitors arrive every day. The place is a hive of activity, accentuated by the shouts of the fishmongers and the roar of trucks, all of it punctuated by the whine of bandsaws. Giant tuna which arrive from the fishing fleet still frozen are sectioned with a bandsaw before the fish is sold to the various retail outlets and restaurants. Large fresh fish, such as tuna and swordfish, are cut up with special, enormous knives with a long blade. Often it requires two people to wield the knife.

Tsukiji is the biggest fish market in the world, selling more than 400 different varieties of fish. The total volume of trade every day amounts to more than 2,000 metric tons of marine products, with a value of about $20,000,000 U.S.

Just outside the market halls lies a veritable El Dorado for sushi-freaks – a row of restaurants and sushi bars, some so tiny and closely packed that one can barely squeeze in through the doors. Food is consumed either standing up or perched on a stool squeezed in between the other sushi eaters at the bar. Here one finds the freshest fish imaginable, newly fetched from the market next door. Of course, we could not resist temptation. Sushi and *sashimi* are available from the start of the day. We soaked up the hectic atmosphere while watching the merchants who were savouring a bowl of warm soup and a cup of green tea.

On the parking lot are a number of small enterprises which peddle vegetables and flowers. With the help of gestures and yen I managed to buy some *wasabi* plants of two different sizes and qualities.

The plants were brought back home, carefully wrapped in damp paper. Fresh *wasabi* is not available where I live in Denmark, so the prospect of grating my own supply instead of using *wasabi* powder filled me with excitement.

Once back in the outer market, we ran into a row of little stores with an open front, displaying *katsuobushi*, fine shavings of dried bonito (*katsuo*), a fish related to mackerel and tuna. The fillets are cooked, salted, smoked, fermented, and dried, becoming rock-hard lumps which keep indefinitely. Before use, the pieces are shaved into paper thin flakes using a special tool, which is a wooden box with a plane-like blade. The flakes are ubiquitous in Japanese cuisine, sprinkled as 'dancing' fish flakes on salad, *tofu*, or rice, and as a main ingredient in *dashi*, the broth which is the base for most Japanese soups.

On the way to the nearest subway station, we again gave in to temptation and visited one of the many well-stocked cookware establishments. Without giving a thought to how much space I had in my suitcases, I was easily pursuaded to buy a stack of the traditional gold patterned, reddish brown lacquerware bowls for *chirashi*-zushi.

Cooked giant shrimp with a clearly visible cross-striped musculature in the tail.

Schematic illustration of a shrimp. The dark sac with the long tail is a digestive gland which produces *enzymes* that quickly break down the muscle tissue when the shrimp is dead. The long tail of this gland shows up as a black vein in the cooked shrimp. This vein must be removed before the shrimp can be used for sushi or *sashimi*.

学
術 SWEET SHRIMP. Shellfish contain more *glycogen* in their muscles than fish, which is why shrimp and lobsters have a sweet taste. Unfortunately, the taste fades at the same rate as the glycogen is converted to *lactic acid*. In a sushi specialty appropriately called sweet shrimp, *amaebi*, the crustacean is eaten raw.

THE TEXTURE OF CRUSTACEANS

Crustaceans have a preponderance of fast muscles and their flesh is white. As they contain more connective tissue than the muscles of bony fish, they are also less tender and dry out more readily.

Typically, crustaceans are about 80% water and have only a little fat, about 2%.

HOW DO CRUSTACEANS TASTE?

Crustaceans contain large amounts of MSG, as well as other pleasant tasting *amino acids,* such as *glycine*. They taste sweet and often have a nutty flavour after they have been cooked. This flavour is brought about when the amino acids bind with sugars to form substances such as *pyrazines* and *thiazoles*. This is the same chemical reaction as the one which is responsible for the browning of meat.

On the other hand, crustaceans have only a little *trimethylamine oxide,* which is why they give off much less 'fish odour' after they have died. But as their *enzymes* are aggressive, they quickly decompose the organism's tissue if the crustaceans are not either frozen or cooked soon after they have been caught.

THE TEXTURE OF MOLLUSCS

Bivalves have two hard shells that need to be able to open to take in food and to close to keep out enemies. The adductor muscle (or muscles, as in some species there are two) needs both to be able to shut quickly and to hold the shells together tightly for long periods of time. Hence, they are composed of both fast and slow muscle fibres. The fast muscle fibres are white, translucent, and soft like white fish muscles. The slow muscle fibres need to be incredibly strong. They do not make use of *myoglobin* as do the slow muscles in fish, but instead perform their function using an interlocking mechanism that expends very little energy to keep the shell closed once it is in that position. These unusually strong muscles, which contain a great deal of connective tissue, are glassy and closely resemble chicken cartilage. It is mostly the fast, softer adductor muscle which is edible.

In the case of some bivalves, for example, oysters, one eats not just the adductor muscle, but also the rest of the innards, whereas the adductor muscle is the only part of a scallop that is used for sushi.

Cephalopods, such as octopuses and cuttlefish, have longer muscle fibres and more connective tissue than bony fish. Apart from being more abundant, the muscle fibres are also constructed in a more complex manner. The individual muscle fibre has a diameter of about 0.004 mm, which is a tenth of the width of that in a fish. Because the fibres are so fine, cephalopod muscles have a smoother structure than those of fish.

The muscle fibres in cephalopods are bound together by connective tissue that contains *collagen*, just as in fish, but there is much more of it. In addition, it is *crosslinked,* which gives it far greater strength, but also makes it much tougher.

Illustration of the structure of a scallop showing the large, edible adductor muscle, which is almost pure white.

学術 **SHELLFISH** such as mussels and oysters are comprised of about 80% water and only a small quantity of *fats*, typically 0.1-1.2%.

学術 **CROSSLINKING OF FIBRES** is a way to make soft material more robust and tough. The chemical explanation is that strong links are formed in all directions among the long-chain *polymers*, for example, *proteins*, which make up the soft tissue.

An example of an industrial application that takes advantage of crosslinking is the conversion of the polymer polyisoprene to rubber by a process called vulcanization. The result is a strong material with desirable elastic properties.

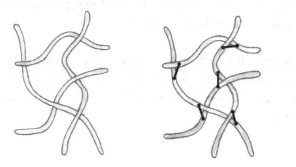

Crosslinking in a network of long molecules or fibres.

How do molluscs taste?

Just like salt water fish, *bivalves* living in the ocean can counteract the *osmotic* effect of salinity by increasing the concentration of certain substances in their cells. Bivalves contain a particularly large proportion of amino acids, among them *glycine* and *glutamic acid*, as well as alanine, proline, and arginine. In addition, the *glycogen* content used to fuel their fast muscles serves to emphasize the sweet-spicy taste of molluscs.

Cephalopods, for example, octopuses and squid, utilize *trimethylamine oxide* to a greater extent than do bivalves to maintain an osmotic balance in relation to the surrounding salt water. As trimethylamine oxide has no taste and these cephalopods contain smaller quantities of the sweetish *amino acids*, they have a less sweet and spicy taste than bivalves.

When a mollusc dies, the trimethylamine oxide in the muscles is converted, with the help of the organism's own very active *enzymes*, to *trimethylamine*, which quickly results in an offensive 'fish odour'.

The sea urchin – an echinoderm

Sea urchins have a hard shell which is reinforced with calcium plates and armoured with sharp needles. The needles are unique in that they are single crystals of the mineral *calcite*, which is overwhelmingly made up of *calcium carbonate* (lime).

The reproductive organs of sea urchins – testicles and ovaries, which can be difficult to distinguish from each other – make up two thirds of the inner parts of the organism. They have a yellowish brown colour and a shape that resembles a walnut or a miniature brain. These are the only parts of the sea urchin which are edible and are used for sushi. In this context, the reproductive organs are referred to as sea urchin roe (*uni*).

The relatively high *fat* content and the large quantities of *amino acids* and *inosine monophosphate* contribute to the strong taste of sea urchin. The roe also has an intense flavour of sea water and *bromophenols*.

Picture of the cross-section of a sea urchin on which one can see the yellowish brown roe (*uni*).

AQUACULTURE OF FISH AND SHELLFISH

Over-fishing has put an increasing strain on the resources of the oceans and is a threat to many fish and shellfish populations. Some species, which formerly were very common, have disappeared or become much less abundant, cod and herring being prime examples. It is estimated that up to 70% of all species of fish are in the process of being depleted. Availability of fresh fish and shellfish increasingly depends on their cultivation in special marine farms. Hence, the production of seafood under controlled conditions solves some of the problems related to environmental sustainability.

Aquaculture has some clear-cut advantages, among them quality control, rapid growth, and assured, stable supplies. In addition, there are no complications arising from the incidental catch of other species and it is possible to process and cool the harvest very quickly. Finally, it is obvious that the cultivation of fish is the most environmentally friendly way of ensuring our need for animal proteins and essential fats.

On the other hand, there are a number of drawbacks associated with the consumption of farmed fish. First, there may be problems with interbreeding of introduced species with wild stock. Secondly, even if they often have a greater fat content than wild fish and, therefore, have a softer texture, farmed fish can taste rather bland. This is due to the limited opportunity for free movement in the holding nets, which causes the fish muscles to have a less firm texture. Normally, farmed fish are less rich in *omega-3 fats* than their wild counterparts. There are also significant problems with water pollution resulting from the feeds used, the accumulation of waste products from the fish themselves, and the *antibiotics* employed to keep the stock disease free. Some of these problems can be attenuated by combining fish farming with the farming of seaweed and shellfish.

Among crustaceans, it is especially prawns and shrimp which are cultivated. A third of the world's production of shrimp is now based on aquaculture.

Of the bivalves, immobile mussels, which clamp themselves onto rocks and the seabed, are especially suitable for aquaculture. They are grown on ropes and nets that are suspended in the water.

学術 AQUACULTURE. Globally, about half of the world's seafood is farmed commercially and over eighty different species are cultivated. For example, more than a third of warm water shrimp and of salmon for domestic consumption are now sourced from marine farms.

学術 FRESH FISH FROM THE FARM – sort of. Cultivated salt water fish are lacking in *bromophenols* and, therefore, do not have the same fresh smell of the sea which we associate with fresh fish.

学術 **Salmon or tuna?** Salmon is situated lower in the food chain than tuna, which not only has a long life span but also ranges widely in the seas of the world. Hence, salmon usually has a smaller build-up of toxins than tuna.

学術 **Shellfish poisoning.** Mussels, oysters, crabs, and other shellfish sometimes accumulate *toxins* because they filter feed on one-celled algae called dinoflagellates and diatoms. Toxins are the primary defence mechanism of these microorganisms and, as the toxins are heat-stable, cooking does not render them harmless. That is why one should avoid eating shellfish if one suspects that it is contaminated; doing so can result in what is called shellfish poisoning. Fish with fins normally do not contain these toxins.

Environmental toxins in fish

Environmental toxins, notably heavy metals, *dioxins*, *PCB*s, and pesticides tend to accumulate in the food chains of the oceans. Smaller organisms, like crustaceans and especially shellfish, which live on the sea bed, take in poisons such as mercury, which are then passed on to the larger predator fish. Because heavy metals are difficult to excrete, they gradually become more concentrated at the top of the food chain. Tuna, swordfish, sharks, and other fish that have a long life span and range over large areas of the ocean end up with significant concentrations of toxins. As many of the organic pollutants dissolve very readily in *fats*, it is particularly the oily fish which are affected.

There is much controversy about the extent to which the heavy metal content of edible fish is a problem, as well as the effect these pollutants have on our health. This debate is closely related to the discussion about whether breastfeeding of infants poses health risks, with some claiming that mother's milk contains too many environmental toxins.

Recent research seems to indicate that the threat to health posed by heavy metals might have been overestimated and that one should not avoid eating fish solely for this reason.

Larva of a herring worm, *Anisakis simplex*, which is a naturally occurring parasite in such fish as mackerel, herring, cod, and cuttlefish.

PARASITES IN FISH

Parasites are neither bacteria nor viruses. Rather they are animals of varying sizes, ranging from the microscopic, unicellular *protozoa* to multicellular worms. For a part of their life cycle, they inhabit a host, for example, humans. They are found naturally in some types of raw fish and shellfish and can be transferred to us when we eat these, allowing parasites to take up residence in our muscle tissue. Under no circumstances should one eat raw fish or shellfish if there is the least suspicion that they might be infested with parasites.

Fortunately, appropriate handling of the fish can destroy the parasites. In contrast to bacteria and viruses, parasites cannot withstand temperatures in excess of 60°C or lower than -35°C (a temperature lower than that to which a common household freezer can be set). It is best to eliminate them by marinating the fish in salt and vinegar.

Parasites like *anisakis* (popularly known as the herring worm) can occur in mackerel, herring, cod, salmon, and cuttlefish. The danger posed by parasites is greater in fresh water fish than in those that live in very salty water. Serious infections can result from flatworms in fresh water fish and in crabs and crayfish which have lived in brackish water. On the other hand, tuna are very rarely affected by parasites and farmed salmon have a much lower incidence of infestation than their wild counterparts.

学術 **IS IT DANGEROUS** to eat raw fish? Raw fish or fish that is undercooked or not sufficiently marinated can contain parasites, but not all raw fish is infested. If you are worried that a fish which you want to use to make sushi has parasites, you can freeze it for 24 hours at the lowest possible temperature. An alternative is to marinate the fish in salt and vinegar. In some countries, restaurants that serve raw fish must routinely freeze the fish to a temperature of at least -20°C for a period of at least 24 hours.

Is it dangerous to eat fish?

ABOUT BALANCING THE GOOD AND THE BAD, AND HOW WE
CAN PRODUCE MORE FISH FOR A HUNGRY WORLD

We know that a diet which is rich in fresh fish and shellfish and, therefore, in omega-3 fats contributes to reducing the incidence of cardiovascular disease. In addition, there is a growing awareness that modern eating patterns with too few omega-3 fats and too many omega-6 fats are partly to blame for the rapid increase in the Western world of mental illness and neurodegenerative diseases such as Alzheimer's disease. Fish contain special fats and minerals which counteract these diseases and which are vital elements in brain and nervous system formation and function. Public health officials also recommend that we should eat fish more regularly than we do now.

But the problem is that fish can contain substances which are harmful to the health of the consumer. Fish, especially those which are oily as well as those which are high up in the food chain like tuna, can be contaminated with mercury, pesticides, and PCBs. About 30% of the mercury pollution in our environment is attributed to natural causes, for example, volcanic eruptions, but 70% is due to human activity, especially emissions from power plants burning high sulphur coal and garbage incinerators. All mercury compounds act on the nervous systems as dangerous poisons. Mercury in the form of methylmercury has been under intense scrutiny following two catastrophic instances of mercury poisoning, one related to eating contaminated fish in Japan in the 1950's and the other to using wheat grains treated with methylmercury fungicide in Iraq in the 1970's. Healthy mothers gave birth to children who suffered from severe psychological and neurological defects.

How much of a problem is there with fish and how do we reconcile the health promoting aspects of fish consumption with those which are potentially damaging? The general public is often improperly informed, or even misinformed, about eating fish or misunderstands official communications on the subject. This has led to a great deal of confusion and anxiety. The question is: do we compound the damage by making people unnecessarily afraid of including fish in their diet, thereby taking away their most important source of omega-3 fats? Where is the balance between the two? What is needed to shed light on this are studies of population groups, especially children, who had fetal exposure to methylmercury because their mothers ate fish during a vital phase of the formation and development of the brain and central nervous system of the fetus. But such research is exceptionally difficult to undertake.

There are three research projects which are worthy of attention. The first was undertaken on the Faero Islands because the local inhabitants eat so much fish and whale meat that their exposure to contaminants in their diet is about five times that of most people in the West. Designing this study to produce meaningful results for methylmercury was very complicated because other factors come into play, given that whale meat bioaccumulates a whole series of other environmental toxins such as PCBs and pesticides. Nevertheless, the principal conclusion was that mercury pollution causes measurable, if small, effects on the neurological development of children.

This result seemingly contradicts those derived from another study carried out in the Seychelles in the Indian Ocean using a subject population that is considered ideal because one can isolate the potential effects of methylmercury. The research is deemed to be especially useful for a number of reasons. Firstly, it covers a long period of time. Secondly, the fish caught in these waters are contaminated only by methylmercury. Thirdly, the average consumption of between eight and twelve servings of fish per week is substantial. The main conclusion was that there are no demonstrable disadvantages to an intake which is up to fifty times as great as that of the average North American. It would actually appear that children born to these mothers, who ate fish in large quantities, have better visual and cognitive abilities. These conclusions are supported by a recent study of about 12,000 pregnant women in England and their children. No adverse effects have been detected even when the mothers had a very large intake of fish during pregnancy.

Present knowledge clearly leads to the conclusion that the advantages of including fish in the diet easily outweigh the drawbacks. Hence, there is no reason to discourage North Americans and Europeans from increasing their consumption of fish and shellfish, even though there is still a lingering unanswered question concerning what the recommended weekly intake should be. Many experts are in agreement that there is too much emphasis, not least in the popular media, on the risks instead of on the actual healthful benefits. Even though we know that population groups who eat a great deal of fish are generally healthier, live longer, have a lower incidence of cardiovascular disease, and are possibly more intelligent in the bargain, we seem to hear only the negative reports. People have a neurotic relationship with their food and they would rather avoid foodstuffs with a low or hypothetical risk of being harmful than take their chances on ending up in the cardiac ward.

But this leads to another problem. If the recommendation is that we should eat more fish, possibly twice as much, where will the supply come from? Globally, the catch is either stagnant or decreasing and there are serious concerns about the viability of some fish stocks. It is estimated that about half of the world's species of edible fish are being used to capacity, about a quarter are being over-fished, and only the remainder are not fully harvested.

Given that the per capita consumption is approximately constant and that the world's population is increasing, it is already apparent that there will be scarcities. It is expected that by 2010 we will have a global shortfall of up to 10 million metric tons of fish, growing to 50-80 metric tons by 2030.

In view of the fact that farmed fish presently account for about 50% of the total catch, one might think that the problem could be rectified by a greater resort to aquaculture. But additional production from this source will have difficulty in making up the deficits because the current output is increasing by just a few percentage points annually and the increase is no match for the demand created by population growth. It is interesting that 90% of the farmed fish originate in Asia, especially China, and the largest proportion of this supply is destined for the European Union, Japan, and the United States. The conclusion to be drawn seems to be that without a completely different approach to the administration of the oceans' bounty, there will simply not be enough fish to go around.

Experts in the field of nutrition have predicted that the recommendations from public bodies that we will see in the future will inevitably come to mirror available food supplies. The upshot of this is that if we cannot obtain an adequate supply of fish, it is doubtful that the authorities will recommend that we increase their proportion in our diet, despite the known health benefits that would underpin such an admonition. One thing is certain – our children and grandchildren will end up eating foods that are different from the ones we now consume.

'Plants from the sea'

Seaweed are algae that are found in all climate zones all around the world. The smallest algae are unicellular and some are related to bacteria and fungi. The largest are multicellular seaweeds that are several meters long. Most algae are able to photosynthesize, but they are not real plants. Seaweed is a neglected nutritional source in the Western world, but is fundamental in Japanese cuisine, especially in sushi.

ALGAE AND SEAWEED

Algae encompass a large group of very different organisms, of which many live in aquatic environments. There are at least 35,000 different species in existence at present.

Seaweed is a catch-all designation for the larger algae, the macro-algae, which are not real plants although that is what they are often called. There are about 10,000 known species. Macro-algae are classified as either green, red, or brown algae and most of them are edible.

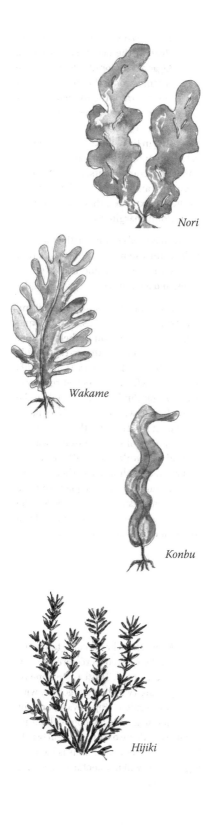

Nori

Seaweed does not have leaves in the botanical sense of the word, but instead has stems and flat filaments, referred to as fronds. As they have no roots, some species attach themselves to rocks or the seabed while others float freely in the water.

Seaweed is an important food source in Asia, especially China, Japan, and Korea. In Japan seaweed accounts for up to 10% of the population's nutritional intake. Japanese cuisine avails itself of an abundance of seaweed, especially *nori*, *wakame*, *konbu*, and *hijiki*.

Wakame

Nori is produced by making paper-thin sheets of a red algae of the family *Porphyra*. *Nori* is used to make *maki* rolls. *Wakame* is a brown algae that has a mild taste and is often used in soups and salads. *Konbu* is a brown algae with large lamellar fronds, known as kelp in English. It is rich in MSG (*monosodium glutamate*), which is the basis for the 'fifth taste', *umami*. Hence, *konbu* is often used for making soup stock. *Hijiki* is a brown seaweed with long, thin dark green fronds. It is often simmered and used as a condiment or in salads.

Konbu

THE TASTE OF HEALTHY SEAWEED

Because seaweed must live in saline water, different substances are concentrated in the cells of the seaweed to maintain the *osmotic* balance. Many of these substances contribute to the characteristic taste of seaweed, an example of this being *mannitol*, a sugar alcohol named after its resemblance to the taste of the Biblical manna, which imparts a characteristic sweet taste. Up to a quarter of the dry weight can be comprised of mannitol. This substance, which is also found in fungi such as edible mushrooms, cannot be metabo-

Hijiki

学術 The biomass of seaweed and algae in the ocean is enormous. Algae are responsible for 80% of the organic production on our planet and almost 90% of the oxygen production. Seaweed regenerates very quickly, making it a sustainable crop that can be harvested in the wild as well as be cultivated. It contains a long list of valuable nutritional components, minerals, trace elements, and *vitamins*. The Western world has exploited this important resource only marginally.

Seaweed is used commercially in animal fodder, for the production of gelation agents, in cosmetics and herbal medicines, and as fertilizer.

Except as emulsifiers and thickening agents in prepared foods, seaweed and algae are not common in the daily diet of most Westeners. This is regrettable because they are healthy, tasty foodstuffs and a shamefully overlooked resource. Many people think of seaweed only as the brown bladderwrack commonly found on beaches at low tide. But there are many other species of seaweed in the seas all around the world, especially in waters that are more saline. All of them are edible, although not all are equally suitable for human consumption. Of course, one should be cautious and only eat those found in uncontaminated areas.

学術 *Nori* which has become too soggy can be roasted to restore some of its crispness. This can be done by placing two sheets together, with the smooth side facing inward, and for a brief moment (2 to 3 seconds), holding them over a flame or placing them briefly in a toaster oven. While this is happening, the *nori* will momentarily take on a greener hue.

lized by humans and consequently has few calories. Other taste and olfactory substances found in seaweed are the *amino acid glutamic acid* and certain sulphur compounds (e.g., *dimethyl sulphide*).

Dried seaweed can be stored for a very long time in a sealed package. It has a salty and slightly spicy taste and, like a fresh sea breeze, smells of *bromophenol* and iodine.

Many species of seaweed are rich in *vitamins* A, B, C, and E, in addition to minerals and, in particular, iodine. There is up to ten times as much mineral content in seaweed as in terrestrial plants.

Some species of seaweed, such as *wakame* and *konbu*, are thick and chewy, requiring cooking or roasting to make them edible. Others, such as the red alga *nori* (*Porphyra yezoensis*) and the green alga known as sea lettuce (*Ulva lactuca*), are very delicate, having very thin fronds which can be eaten as is. The fronds are no thicker than one or two layers of cells.

Seaweed and algae contain the superunsaturated *omega-3 fatty acids*, especially *eicosapentaenoic acid* (*EPA*), and together with phytoplankton are the source of these essential *fats* for fish and shellfish.

Nori – thin sheets of seaweed for making *maki*-zushi

The red alga *nori* (*Porphyra yezoensis*) is farmed and harvested to produce the paper-thin sheets of *nori*, which have a slightly smoked flavour with only a trace of iodine. The sheets are one of the most important ingredients for preparing sushi. They are commonly used to make the well-known standard sushi *maki* rolls and handrolls (*temaki*-zushi), to wrap around individual rice balls topped with fish or, more often, roe to make battleship sushi (*gunkan-maki*). Cut into fine strips or tiny pieces, *nori* is also sprinkled on food (as *furikake*) and used as a taste additive in rice, salads, and soups.

After the red alga fronds are collected or harvested, they are rinsed and chopped up to produce a kind of pulp from which a paper-thin layer is spread on a bamboo mat and subsequently dried, causing the colour to change from red to green or black. The process resembles that used to make paper. The resulting sheets, called *hoshi-nori*, are roasted, dried, and cut to a standard size, 17.5 cm × 22.5 cm, and

珍
談
The Roman poet Virgil apparently said that there is nothing more disgusting than seaweed. It is true that seaweed smells awful when it rots, producing a stinky gas, *dimethyl sulphide*. Actually seaweed and phytoplankton are responsible for the bulk of the *dimethyl sulphide* that is released into the atmosphere. Because this gas plays an important role in cloud formation over the oceans, which acts as a cooling mechanism, seaweed and phytoplankton play a role in regulating the temperature of the earth.

学
術
Nori from Japan is generally the best there is for making sushi, but good *nori* is also being produced in other countries, notably China and Korea. Three different stretches of coastline in Japan are singled out for the quality of their *nori*: Tokyo Bay (Chiba), Kobe Bay (Hyogo), and Ariaka Sea (Saga). Sheets of *nori* from Kobe are usually thicker than the others and have a stronger taste. That from Saga is very smooth and mossy-green with dark spots. Differences in the quality of *nori* can result in price differentials with a factor of fifty between the cheapest and the most expensive.

then packaged in bundles, usually five or ten sheets. As roasted *nori*, also called *yaki-nori*, absorbs water, the sheets must be stored in tightly sealed packages, often with a small package of silica gel to absorb moisture. It is recommended that *nori* should be stored in a dark place.

Salt and flavourings such as soy sauce and sesame oil can be added to *nori* before it is roasted. In this form it is often used for snacks, folded around a rice ball, or as *yaki-nori* (*ajitsuke nori*). The latter are chopped up as small pieces of roasted flavoured *nori* sheets that can be sprinkled (as *furikake*) on rice and salads.

There is a great variation in the quality, thickness, and colour of *nori* sheets sold for making sushi and the price is adjusted accordingly. The sheets are usually glossy and smooth on one side and matte and rougher on the other. On the dull side one can make out the impression left by the bamboo mat on which the seaweed leaves were dried. The smoother and the darker the sheets, the better the quality. Very uniform, thin, dense sheets are the finest and most expensive. Thicker ones with many holes are the cheapest. *Nori* sheets with a reddish sheen which are matte on both sides are considered to be of poor quality.

Nori sheets of three different qualities. The one on the right is the best – finer and darker green.

An abundance of *nori* thanks to 'The Mother of the Sea'

Every year, about 350,000 wet metric tons of red algae of the genus *Porphyra*, used to make *nori*, are harvested in Japan. *Porphyra* is known commonly as purple laver because it has thin, translucent reddish fronds. Red algae are farmed in Japan on nets in the sea and over 60,000 hectares of Japanese coastline are given over to this form of aquaculture. This, together with the amount produced in China, makes *Porphyra* the world's biggest single marine crop.

Large-scale commercial farming of seaweed only became possible in the 1950's. The Japanese started to cultivate seaweed as far back as the 17th Century, but the output was limited and the enterprise was fraught with difficulty. The problem was that there was not enough knowledge of the biology of *Porphyra* to be able to find effective methods of growing it. Consequently, *nori* was a rare and an expensive commodity. In 1949, the British researcher Kathleen Mary Drew-Baker (1901-1957) discovered that *Porphyra* actually has a complicated life history with several different stages. During one of these stages, the alga spores burrow into the pores and crevices of the shells of bivalves. Earlier attempts to farm red algae had failed, because the seashells necessary to support this part of the cycle were missing from the places where they were being raised.

Drew-Baker's discovery is a wonderful example of how basic science research can translate into knowledge with a practical application. In this case, it laid the basis for a type of aquaculture that has made a meaningful difference in the yield of large quantities of healthy foodstuffs. It is not without reason that the Japanese celebrate her as 'The Mother of the Sea' and stage a festival in her honour every year on the 14th of April.

Celebration of the Drew-Baker festival on April 14 at the banks of the Ariake Sea in the southern part of Japan.

SOYBEANS: *TOFU*, *SHŌYU*, AND *MISO*

Soybeans are full of proteins and unsaturated fats, which make them an important nutritional source for both humans and animals. The beans can be eaten either as they are (*edamame*) or processed into *tofu*, *shōyu* (soy sauce), *miso*, or a whole series of other items. Soy sauce is the most common flavouring agent in the Japanese cuisine. Soy products are also an essential part of every Japanese meal, especially sushi.

PROTEINS AND FATS IN SOYBEANS

Soybeans (*daizu*) are very rich in *proteins*, generally twice the amount that is found in other vegetables. The dry weight composition of soybeans is as follows: 41% proteins, 20% *fats*, 8% *carbohydrates*, and the balance fibre and minerals. This makes them an important source of nutrition for both humans and animals. There is great variety in the type of soybeans cultivated, either to be eaten as beans or processed into *tofu, shōyu* (soy sauce), *miso,* and an abundance of other products.

The beans contain large quantities of polyunsaturated *fatty acids*, which are easily *oxidized* and broken down by the beans' own *enzymes* when they are crushed. The resulting substances are made up of short chain *carbohydrates*, which have an aroma that we normally associate with the taste of beans.

Tofu

Tofu is produced from soy milk, itself made from ground soybeans which have first been soaked in water. The milk, which is pasteurized at high temperatures, is rich in calcium and vitamins. *Calcium sulphate* (gypsum), added when the milk is at 70-80°C, causes it to form into a solid mass, known as a *gel*. This process *coagulates* the soy proteins.

The solid mass is drained on a filter, lightly pressed together, and then washed. The end result is *tofu*, which is about 90% water. The dry matter is made up of about 55% proteins and 22% fats.

学術 BEANS ARE GOOD FOR YOU! Apart from being rich in proteins, soybeans (*Glycine maximus*) are an important source of unsaturated *fatty acids*, especially *oleic acid* (22%), *omega-6 linoleic acid* (53%), and *omega-3 alpha-linolenic acid* (8%). Saturated fatty acids make up about 11% of the net weight. (Source: Belitz et al. (2004), Table 14.11, p. 653)

学術 PROTEIN COAGULATION. *Proteins* are molecules which are often water soluble. The capacity of the proteins to bind together is dependent on their surface area and, possibly, their electric charge, as well as what else is in the water. By adding appropriate salts to the water, one can control the strength of the electrical interaction between the proteins, forcing them to bind together. This is known as protein *coagulation*.

When a salt such as *calcium sulphate*, $CaSO_4$, is added to the suspension of proteins in soy milk, the Ca^{++} ions bind to the proteins, which in turn bind to each other. In Japan, another salt called *nigari* has traditionally been used as the coagulant in *tofu* production. This is bittern or *magnesium chloride*, $MgCl_2$, derived from sea salt. During the coagulation process, the magnesium ions, Mg^{++}, act in the same way as the Ca^{++} ions.

A well-known example of coagulation is the formation of cheese curds, which occurs when the milk protein *casein* forms clumps when milk goes sour, either naturally or when acidified.

珍
談
It is said that soy sauce was discovered when a worker lay down to rest under a vat in which soybeans were being fermented to make *miso*. He tasted the liquid which was seeping out through the cracks in the vat and discovered that it was delicious.

学
術
Kōji is made from rice, barley, or soybeans seeded with the mold *Aspergillus oryzae*. It is used to initiate the *fermentation* process which results in soy sauce, *miso, sake, mirin*, or *shōchū*.

Shōyu (soy sauce)

Soy sauce is not really a sauce in the true sense of the word; it is more of a taste enhancer. It is produced using a sophisticated, drawn out *fermentation* process that starts with cooked soybeans either mixed with wheat (*shōyu*) or without wheat (*tamari*).

The first step is the production of a *fermentation* starter (*kōji*), which is somewhat like a sourdough, made of polished rice, barley, or soybeans seeded with the mold *Aspergillus oryzae*. The spores of this fungus sprout and form a *mycelium*, which produces *enzymes* that can break down the *proteins* and *fats* in the soybeans.

After about two days, the starter is introduced into the soybean mass together with saline water to initiate the fermentation process. The salt causes the fungus to die, but its enzymes remain active in the mixture, which has a low oxygen content. At the same time, yeast and salt-loving bacteria, especially *lactic acid bacteria*, get to work, resulting in the production of a number of taste and aromatic substances, among them *amino acids* such as *glutamic acid*.

The result of this process is *miso*, a soft solid with the consistency of paste, and a left-over aqueous phase. The latter is the original type of soy sauce, also known as *tamari*, made without the addition of wheat.

Currently, Japanese soy sauce is produced using a process introduced in the 17th Century. Both wheat and soybeans are used in different proportions, each contributing to the distinctive taste of *shōyu*.

Japanese soy sauce – *shōyu*

The wheat is roasted and ground and helps to impart a sweet taste and greater alcohol (*ethanol*) content to the soy sauce. Fermentation takes a long time, typically half a year, and is carried out at 15-30°C. Working in unison, the bacteria and the yeast create hundreds of different taste and aromatic substances, which also react with each other. Those which particularly contribute to the characteristic taste of *shōyu* are *pyrazines*, *amino acids* (especially *glutamic acid*), *alcohol*, *esters*, and compounds of amino acids and *sugars* (resulting from *Maillard reactions*). Good quality soy sauce should not have any sediment.

Miso

Miso is produced as a paste by a *fermentation* process initiated by the seeded starter *kōji*. The paste can be made from either soybeans, rice, or barley.

The fermentation is carried out in cedar barrels at a temperature of 30-40°C and can last from a few months to a year. This lengthy process is necessary to produce the taste substances which are characteristic of *miso*. A faster process is now used for the commercial production of *miso*, and artificial taste enhancers are added to compensate for the rapid fermentation.

Miso paste is used as a medium, *miso-zuke*, to preserve vegetables, for example, radishes (*daikon*).

Soup with *miso* is a regular component of every Japanese meal, including breakfast.

学術 THE COLOUR OF *MISO* is dependent on the ingredients used and the length of the *fermentation* process. A large proportion of wheat results in a *miso* that is lighter in colour, and longer *fermentation* results in a darker one.

The so-called white *miso*, *shiro-miso*, is a specialty from Kyoto. It is very sweet and is used in salad dressings and candies.

Red *miso*, *aka-miso*, has a brown to reddish colour and is made from rice.

学術 *MISO*, together with soy sauce, is the most common flavour additive in Japanese cuisine, a sort of universal spice. The '*mi*' in *miso* means something like aroma. *Miso* is very nutritious on account of its large *protein* content, ca. 14%.

White *miso*

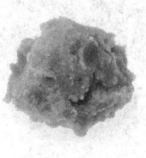

Dark *miso*

Red *miso*

Japanese soy sauce – *SHŌYU*

Always use Japanese soy sauce and not Chinese soy sauce for sushi and *sashimi*. I feel that Chinese soy sauce, which is made exclusively from soybeans, is too dark and has too strong a taste for sushi. Furthermore, Chinese soy sauce, as well as Japanese *tamari*, is cloudy and thus gives an aesthetically less pleasing impression when combined with the clean appearance of the fish and the rice. Once a bottle of Japanese *shōyu* has been opened, it should be stored in the refrigerator.

Shōyu lasts a long time because of its high salt content, but in the course of a few months the contents of an opened bottle will change taste and become dark and murky, to the extent that I myself would avoid using it for sushi. It is still perfectly usable for marinades and in hot dishes. *Shōyu* contains about 12% salt, 8% *amino acids*, 2% *ethanol*, 1% *carbohydrates*, and 1% *lactic acid*. Some varieties are available in reduced salt format.

RICE, RICE WINE, AND RICE VINEGAR

Rice is probably the single most important source of human nutrition. For more than half of the world's people, rice is their staple. There are several thousand varieties of rice, which are roughly divided into three groups: those with short, long, and medium length grains. In Japanese cuisine, and especially in the preparation of sushi, only short-grained rice is used. Rice wine (*sake*) and rice vinegar (*su*) are other important products made from rice.

WHAT IS RICE MADE UP OF?

In contrast to some other Asian culinary arts, for example, that of China, the Japanese kitchen uses almost exclusively polished rice. Milling and polishing remove the outer husk, the rice bran, and most of the rice germ, in the process taking away virtually all the *fats*, *proteins*, and *enzymes*. Only the *starchy* white endosperm remains. For this reason, polished rice can be stored for several years.

Rice destined for rice wine (*sake*) is highly polished, generally losing about 30% of the outer part of the grains during milling. For the finest quality of rice wine, one half of the outer grain is removed before it is cooked into a mash.

The endosperm of the grain of rice is built up of granules containing starch, which are a few thousandths of a millimetre in size. The starch consists of two types of *polysaccharides* (*carbohydrates*): *amylose* and *amylopectin*. In short-grained rice there is proportionately more amylose than in long-grained rice. Rice does not have any *gluten* and can, therefore, be eaten by those who are gluten intolerant.

In a dry grain of rice the starch molecules in the starch granules are tightly packed, as in a crystal. Each of the starch granules is surrounded by a coat of proteins. That is why the secret of cooking rice, for example, to make sushi, lies precisely in taking advantage of the properties of these starch granules in relation to water.

RICE COOKING

Rice starch granules contain very little water and, when cold, they are able only slowly and to a limited degree to absorb water. But when heated, the crystalline *starch* will start to melt at temperatures of over ca. 60-70°C while it absorbs water. As the protein coating on the surface of the starch granules is difficult to dissolve in water, the coating serves to ensure that the rice grains maintain their shape. In the meanwhile, the proteins on the individual granules mesh into each other and keep the starch granules glued together.

Starch can absorb large quantities of water, which naturally is the reason why ground starch can be used to thicken sauces. Exposure to liquid causes the rice granules to swell, and the starch *gelatinates* and forms what is technically referred to by the term *gel*. Cooling can only partially reverse this process.

A kernel of rice consists, like other cereals, of a *starchy* inner part (endosperm), a layer of bran, and an outer husk. In addition, there is a rice germ (the oval shape with the dot seen at the lower right of the illustration). Polished rice, which retains only the endosperm, is used in the preparation of sushi.

珍
談 **IT IS SAID THAT** rice symbolizes life and the meaning of everything. According to the *Shinto* religion, the Japanese emperor is the living embodiment of *Ninigo-no-mikoto*, the god of the ripened rice plant.

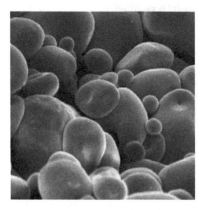

Starch granules in rice. Each of them is about 0.003-0.008 mm in diameter.

学術 **COOKING RICE IN A STRAW BOX.** It is generally known that one can finish the cooking process by taking the pot off the stove before the rice is fully cooked and putting it in the bed under a comforter or, in days gone by, in a box of straw. The reason is that the process whereby *starch* absorbs water and *gelatinates* is optimized in the temperature range from 70 to 100°C. Therefore, if the pot is brought to the boiling point and then well insulated, it is possible to maintain this temperature interval and the rice can finish cooking on its own.

学術 **INCREASED PRESSURE** reduces the cooking time for rice. Water boils at 100°C at normal atmospheric pressure. If one tries to boil water at the top of the Himalayas, one will discover that the boiling point is much lower, at about 80°C. Atmospheric pressure at an elevation of 8,000 metres is lower than at sea level and this causes boiling point depression. If one tries to boil water where atmospheric pressure is greater, the reverse holds true and one finds that a temperature higher than 100°C is required.

One can take advantage of this physical effect to reduce the length of time it takes to cook food by using a pressure cooker. In a pressure cooker, which is basically a pot with a tightly-sealed lid clamped firmly in place, the water vapour being generated under cooking accumulates, increasing the pressure and resulting in boiling point elevation to about 120°C. A rice cooker works in exactly the same way and thus shortens the preparation time.

Short-grained rice, more than other types, retains some of its softness as it cools and releases part of the moisture. The grains of rice easily clump together without falling apart. This is why sushi rice is very different from 'sticky rice' (also commonly called sweet rice), which is almost totally composed of *amylopectin* and which clusters in large clumps because the individual rice grains are broken down during cooking. Cooked sushi rice has a sweetish taste that is modified by the addition of rice vinegar, which is tart.

The texture of the rice is dependent on its age; the older the rice, the more tightly packed the starch molecules. New rice, less than a year old, more readily absorbs water than older rice, which requires a longer cooking time and a little more water.

Sushi chefs prefer rice that is at least one year old for preparing sushi because it is easier to control the texture of cooked older rice.

学術 **RICE AND VITAMINS.** Rice contains an abundance of *vitamin* B_1 and many minerals, but, as they are located in the husks, they are lost in the polishing process. Beri-beri, which is a disease caused by a deficiency of the B_1 protein (*thiamine*), was already common in Asia when it reached epidemic proportions in the 1870's with the introduction of very efficient steam-driven mills for polishing rice. It was only the discovery of the significance of vitamin B_1 at the beginning of the 20th Century that allowed proper measures to be taken to prevent this scourge.

RICE FOR SUSHI

Over 300 different varieties of rice are cultivated in Japan. While a large proportion of the rice sold for sushi in the Western world is from California or Spain, it is considered to be just as good as its Japanese equivalent.

The water content and hardness of rice varies from one cultivar to another. What is most important for sushi rice is that, when cooked, the individual grains become moist, keep their shape, and retain a certain firmness, but still stick to each other. In addition, the rice should remain soft when it has cooled down. Long-grained rice is unsuitable because it is too firm, does not stick together, and becomes hard when cooled.

Rice with a round grain, like the Arborio variety, is also not usable because the individual grains break apart during cooking.

珍談 *SHARI* – the bones of the Buddha. According to legend, the skeletal remains of Buddha were crushed into small pieces in order to be distributed widely as relics. Sushi chefs often call cooked sushi rice *shari*, a reference to these sacred shards. Presumably it points both to the physical resemblance of the individual grains of rice and to a spiritual understanding of rice as the gift of the gods to humankind.

学術 **Sake for young and old.** Once on a flight to Tokyo, an older Japanese businessman cast a nod of approval in my direction when I ordered warm *sake* to go with my meal. He later told me that *sake* should be drunk warm, not chilled, as has become the fashion among young Japanese.

On the other hand, very high quality and well aged *sake* is served cooled.

Cheap *sake* in small bottles and in glasses with a lid on them can be found all over Japan in supermarkets, vending machines, and small stores. These are usually also consumed cold.

学術 **Sake tastes** a bit sweet and a little flat, with undertones of acid and fruit. The taste changes as the *sake* ages, so it must be drunk young. An open bottle should be stored in the refrigerator. The quality of *sake* matches the price. There are both dry (*karakuchi*) and sweet (*amakuchi*) varieties.

Ceremonial *sake* barrels on display outside a *Shinto* temple.

Sake – rice wine, a sacred drink

Sake is wine that is made from rice. In Japan's aristocratic period, from about 700 to 1100, it was considered a sacred drink used in rituals and for ceremonial celebrations. To this day, there is a *sake* drinking rite at weddings and the first *sake* of the year is presented to the gods in large, decorated barrels both outside of, and inside, *Shinto* temples.

The production of *sake* is a complex process that is different from making wine from grapes or brewing beer. For *sake*, the entire rice mash is fermented, whereas for wine and beer the *fermentation* involves only the grape juice or a filtered cereal mash, respectively. For this reason it may be somewhat misleading to refer to *sake* as a wine.

For each new batch of *sake*, a *kōji* consisting of polished rice and a fungus culture is prepared in the same way as it is for the fermentation of soy sauce and *miso*. Cooked rice, which has been acidified with the help of bacterial cultures that produce *lactic acid*, is added to the *kōji*. At this point a pure yeast culture is introduced. The resulting mash, called *moto*, matures in the space of about a month. More cooked rice is added several times, and the fermentation continues for up to a few months.

Fermentation is carried out at low temperatures, typically 10-18°C, and in earlier times all *sake* production took place in the winter. In the process, the yeast converts *starch* in the rice to *ethanol*. By adding cooked rice in several steps, it is possible to attain a very high percentage of alcohol in the finished product.

Shōchū – strong stuff

A rice brandy, *shōchū*, with an alcohol content of up to 45%, is produced by distilling *sake*. The best *sake* is not used for brandy; in fact, the distillate is sometimes made from fermented potatoes, corn, or barley.

Shōchū is used in drinks or in the production of a sweet rice wine, *mirin*, designated for cooking.

Mirin – sweet rice wine

Mirin is used exclusively in the preparation of food. It is a sweet, strong rice wine, made by blending *sake* and *shōchū*. While the mixture cannot be fermented any further because of the already high alcohol content (ca. 14%), the remaining *starch* from the rice can be converted *enzymatically* to *glucose*. This results in a very sweet form of rice wine, which has a light amber colour.

Mirin is used in a variety of ways – in cooked meat and vegetable dishes, when making *tamago* (an omelette used for sushi), or together with *shōchū* as a flavouring.

Su – tart rice vinegar

Rice vinegar, *su*, is produced using a fermentation process in the same way as *sake*, except that the yeast and bacteria now exert their activity in the presence of oxygen. The result is that the *starch* in the rice is not broken down to *alcohol*, but instead is converted into *acetic acid* which is sour. Japanese rice vinegar has a mild and sweet taste, which is less sharp than that of grape-based wine vinegar. It also incorporates some of the same subtle aromatic elements found in *sake* and *miso*.

Su is an important ingredient in the preparation of vinegared rice for sushi, but a rice vinegar powder, *sushinoko*, is sometimes substituted for it. It is a white powder composed of dehydrated rice vinegar, sugar, salt, and flavourings derived from seaweed and dried fish (*katsuobushi*).

A large assortment of other cereals is often used in combination with rice to make *su*. Vinegar that is made solely from rice goes by the name of *yonezu*.

珍談 **Legend has it** that the first *sake* was made by having women in *Shinto* temples chew cooked rice and then spitting the resulting mash into a vat. The *enzymes* in the saliva and naturally occurring bacteria and mold converted the rice gruel to an aromatic, alcoholic drink. This is the origin of the name *kuchikami no sake* (chewed in the mouth *sake*), as *kuchi* means mouth.

Sushinoko – rice vinegar powder

Su – tart rice vinegar

SPICES IN JAPANESE CUISINE

Spices play a much less prominent role in Japanese cuisine than in its European counterparts and in those of other parts of the Orient. Spices are not mixed with Japanese food, but are most commonly used as a condiment, for example, as a topping, a sauce, or simply as a complementary side dish. They are generally derived from aromatic plants like *wasabi*, *shiso*, and sesame seeds or from a variety of vegetables which are pickled in brine.

JAPANESE SPICES

The most common and authentic spices for the preparation of sushi are Japanese horse radish (*wasabi*) and perilla leaves (*shiso*), as well as mixtures of toasted dried seaweed, sesame seeds, and dried fish flakes (*furikake*). One could add to the list the different types of brine pickled vegetables (*tsukemono*), an example of which is pickled ginger (*gari*). *Tsukemono* accompany most Japanese meals.

Miso and soy sauce are probably the most utilized flavourings in Japanese cooking. It is important to be aware that soy sauce, with its distinctive combination of sweet and salty tastes, is used to enhance the taste of other foodstuffs, not to overpower them.

WASABI – JAPANESE HORSERADISH

Wasabi (*Wasabia japonica*) is also known as Japanese horseradish, but it is not actually horseradish. The plant is a perennial herb, native to Japan where it grows wild by mountain streams that have cool and very clean water. It is in the same family as cabbages and can be cultivated. The best quality specimens, highly prized by the Japanese, are very valuable. Once harvested, the fresh plants are stored in running water or are frozen.

For sushi and *sashimi* the thick rhizome of the plant is grated to the consistency of a mash. Grating it generates a pungent smell and the mash takes on a sharp, irritating taste due to the formation of *isothiocyanates*, which are also found in mustard oil. The chemical process is identical to the one that takes place when one crushes mustard seeds or grates an ordinary horseradish root.

Similar chemical substances are found in cabbage and onions. They all contain *glucosinolates*, for example, *sinigrin*. After the cells of the plant have been destroyed by grating or chopping, the glucosinolates are converted to isothiocyanates in the presence of water. The conversion is mediated by an *enzyme* called *thioglucosidase*. The release of isothiocyantes is one of the plant's own chemical defence mechanisms.

The grated *wasabi* can be mixed with a little soy sauce and should be used immediately, as it very quickly loses its strong taste. The taste can be amplified and made more long-lasting by the addition of a little rice vinegar.

Wasabi plant. Its thick rhizome can be grated to a mash and used with sushi or *sashimi*.

Fresh *wasabi* is grated to a pulp on a classical grater (*oroshi-gane*), which is made by gluing shark skin to a wooden board.

珍
談
IT IS SAID THAT a good way to avoid the consequences of an excess of *wasabi* is to breathe through the mouth. That way the extent to which the *isothiocyanates* make their way through the nose into the lower air passages and the lungs is decreased.

学
術
AN IRRITATING TASTE. Irritation on the tongue and in the mouth is a chemical effect that has nothing to do with taste. *Wasabi*, mustard, chili, and garlic all cause this irritating effect, which can seem unpleasant while one is eating. Somewhat paradoxically, the irritating taste enhances the experience of a meal since it provokes the production of saliva. The irritation is due to the fact that the chemicals in these plants damage our cells and give rise to a danger signal – pain – that is registered in the brain. In the plants, of course, these substances serve as a chemical defence mechanism.

The acid in the vinegar decreases the enzymatic activity and helps to slow down the formation of *isothiocyanates*. As they are then oxidized more gradually, the sharp taste is preserved for a longer period of time.

Outside of Japan it is hard to come by fresh, genuine *wasabi*, so one frequently resorts to the most readily available substitute product – a fine, pale green powder (*kona wasabi*) packed in an airtight tin or plastic bag. This powder is not real *wasabi*, but ordinary horseradish to which mustard oil and green food colouring have been added.

This powder is dissolved in a little water or rice vinegar to make a paste, which should rest for 5-10 minutes to release the taste elements. It should then be used right away as it does not keep. *Wasabi* powder benefits from being stored in the freezer in a tightly sealed package, with care being taken to keep moisture out.

While grated *wasabi* can be bought as a paste in a tube, it is my own experience that freshly prepared *wasabi* paste made with powder is often superior to that available in tubes. In a sense, the ersatz *wasabi* is a high-quality entity in its own right and should not be regarded simply as a substitute for the real thing.

By far the majority of sushi restaurants, including those with a reputation for excellence, make use of *wasabi* made from *kona wasabi*. This powder usually will also be the amateur chef's only source of *wasabi*.

Wasabi is rich in *vitamin* C and in addition has certain anti-bacterial, and, hence, antiseptic, properties.

In many ways, *wasabi* is the very soul of sushi. In peculiar and surprising ways the sharp, pungent taste of the *isothiocyanates* enhances the delicate taste of fresh fish and shellfish. Some people think that *wasabi* removes undesirable aftertastes. The piercing taste stimulates the production of saliva and digestive juices and in this way helps to whet the appetite. The tingly and slightly warm taste sensations produced by *wasabi* cry out for more sushi!

The *isothiocyanates* in *wasabi* are soluble in water. It is easy to remove the tingling sensation in the mouth by washing it down with water or green tea. This is not true if you burn your tongue with black pepper or chili. *Piperin* and *capsaicin*, the taste substances in these spices, are not water soluble and bind to the taste buds on the surface of the tongue.

Wasabi is also well suited for balancing the mild and slightly flat taste of other ingredients such as avocado or marinated salmon.

学術 **THE GLOBALIZATION OF *WASABI*.** In recent years, *wasabi* has made inroads into Western taste preferences, for example, as a flavouring additive for potato chips, peas, mayonnaise, frozen desserts, and drinks.

Wasabi powder (*kona wasabi*). This is not real *wasabi*, but a substitute product made from ordinary horseradish to which mustard oil and green food colouring are added. The powder is dissolved in a little water or rice vinegar, resulting in a paste that has a more pronounced green colour. Experiment with adding liquid a bit at a time until you hit on the right consistency. The little mounds of *wasabi* that accompany servings of sushi and *sashimi* have a clay-like and firm consistency. On the other hand, it is advantageous to have a slightly more runny paste to put on the fish used for *nigiri-* or *maki-*zushi.

The leaves of green *shiso* have a nettle-like appearance with tiny fine hairs on the top surface. Whole leaves can be wrapped around a rice ball or a piece of sushi. For a more elegant presentation, a few green *shiso* leaves can be substituted for the little serrated green plastic pieces that are often part of a sushi arrangement.

Red *shiso* leaves can be used in the brine pickling of plums or, when dried and salted, they can be crushed and sprinkled on rice or salads.

Shiso – perilla

The *shiso* plant (*Perilla frutescens*) has generally been known as perilla in English, although its Japanese name is now seen more and more frequently. Its leaves are used in a multitude of ways in Japanese cuisine, being prized for their very special aroma and taste. There are three varieties: green (*ao-jiso*), red/purple (*aka-jiso*), and a hybrid with leaves which are green on top and red/purple on the bottom (*aoaka-jiso*). *Shiso* is a shade-loving annual, which is easy to grow in a greenhouse or outside in the summer. Oddly enough, *shiso* has not yet entered the ranks of the herbs commonly available at greengrocers.

Shiso is a member of the mint family, but its taste is actually more evocative of basil. The taste of the red leaves is particularly spicy and quite sharp.

The green leaves are often used as decoration on a plate of *sashimi*, in *tempura*, or simply folded around a piece of *nigiri*-zushi. For *tempura*, which is batter-fried fish, shellfish, or vegetables, only one side of the *shiso* leaves is coated and, because the leaves are so thin, they are fried for only a brief moment. Green *shiso* leaves can be used whole or cut into thin strips (*sengiri*). Mixed into, or placed on top of, rice balls they make a good side dish. With their slightly minty taste, enhanced by just a hint of basil, they are an excellent partner for vinegared rice.

Red varieties of *shiso* are often used to impart a special magenta hue to foods or for preserving plums (*umeboshi*) or brine pickled vegetables (*tsukemono*). The red colour is due to *anthocyanins*, chemicals also found in other plants such as red cabbage and black currants. As the anthocyanins are very water soluble, they readily release their reddish purple pigment to their surroundings.

In Japan the seeds, flowers, stalks, and berries of the *shiso* plant are also used in soups and sauces. The seeds have a high *omega-3 fat* (*alpha-linolenic acid*) content.

In the United States, red *shiso* can also correctly be referred to as 'beefsteak plant'. Unfortunately, it is often confused with Herbst's bloodleaf (*Iresine herbstii*) which is a member of the ameranthus family, while *shiso*, as already noted, belongs to the mint family.

Dried red *shiso* leaves can be crushed and sprinkled on eggplants or used as a colouring agent and preservative in brine pickling, for example, with plums (*umeboshi*).

Shiso also kills bacteria

Perilla aldehyde, the powerful aromatic found in *shiso,* and *perilla alcohol* are in the same chemical family as *perilla acid.* All three of these perilla substances belong to the general group that chemists call *terpenes,* which have a strong effect on cell membranes. A related terpene is *limonene,* known from the smell of dill, pepper, caraway seeds, and citrus rind.

Because the substances that are derived from *perilla acid* have a strong effect on cell *membranes* (they are *amphiphilic*), they have useful preservative properties. It turns out that these substances are able to destroy the cell walls of bacteria and fungi and thereby help to preserve fruits and vegetables. Crushed red *shiso* leaves are used in the brine pickling (*tsukemono*) of plums, cucumbers, and eggplants. The preservative effect is significant. In my refrigerator, I have some brine pickled plums, prepared with home grown red *shiso,* which are over a year old. They are still as fresh and tasty as ever.

It is thought that *shiso* and *perilla aldehyde* also have certain anti-carcinogenic properties.

Sesame seeds (*goma*)

Seeds from the sesame plant (*Sesamum indicum*) come in several colours. For sushi, the white (golden or light brown) and the black ones are the most commonly used. The seeds contain 50% oil and are especially rich in monounsaturated *oleic acids* and the *omega-6 fatty acid linoleic acid.* In addition, the seeds have significant quantities of *antioxidants,* which is the basis of their long shelf life. The black seeds have a more intense flavour than the white ones, but the latter yield more sesame oil when they are crushed.

Before use, sesame seeds are toasted for a few minutes in a warm pan to bring out their familiar nutty aroma. If they are then crushed or ground, the aroma is intensified.

Sesame seeds are used for *maki*-zushi, for example, with avocado, and are also a component of most of the toppings (*furikake*) that are sprinkled on warm rice and a number of the side dishes which will be described later in this book.

White sesame seeds (*shiro goma*) and black sesame seeds (*kuro goma*).

Furikake – Japanese 'spicy topping'

The Japanese have evolved a whole arsenal of dried 'spice mixtures', known as *furikake*, which are sprinkled over warm rice, vegetable dishes, and warm fish. *Furikake* usually combine toasted dried seaweed, toasted sesame seeds, and flaked, dried fish. The sprinkle made from toasted green seaweed, for example, sea lettuce and related Japanese species, is called *ao-nori*.

A special type of *furikake* (*yukari*) is made up of granulated, toasted red *shiso* with salt.

Furikake based on a mixture of toasted *nori*-flakes and sesame seeds.

Yukari is a *furikake* based on dried red *shiso* (*aka-jiso*).

The brain needs fats

ON THE EVOLUTION OF THE HUMAN BRAIN
AND WHY YOU MUST EAT FISH

Very few people are aware that the human brain consists mostly of fat. About 60% of its net weight is made up of fatty substances called lipids, and about half of these lipids are the polyunsaturated fats, omega-3 and omega-6 fatty acids. These fatty acids are labelled essential because our bodies themselves can generate them to only a very limited extent. Consequently, we have to get them from our food. The majority of the omega-3 and omega-6 fatty acids in the brain are the very long-chain and superunsaturated fatty acids: arachidonic acid (AA), eicosapentaenoic acid (EPA), and docosahexaenoic acid (DHA). Comparably large quantities of superunsaturated fatty acids are found in other parts of the nervous system, especially in the retina, which can actually be considered to be part of the brain. The omega-6 fatty acid AA can be obtained from animal products such as meat, liver, and eggs. But the question is: where do the omega-3 fatty acids EPA and, above all, DHA come from?

It is worth noting that the relatively large proportion of super-unsaturated fatty acids in the brain is not particular to humans. The brains of a long list of other mammals, of reptiles, and of fish have a comparable level of fatty acids. By way of contrast, though, the combinations of fats in the muscles, liver, and other organs of the different animals vary widely from species to species. This universal preponderance of superunsaturated fats in the brain signals that there is something very peculiar about it. And this something has to do with its evolutionary history.

In the course of the past one million years or so, the succession of upright primates who are said to be our ancestors has experienced a substantial growth in the size of the brain. At one point in the sequence, the modern *Homo sapiens* appeared, thought to have originated in Africa between 100,000 and 200,000 years ago. The British neurochemist Michael Crawford has hypothesized that availability of DHA was the determining factor for the evolution of the large brain in humans.

Crawford's point of departure is the observation that it is the presence of a large brain that distinguishes *Homo sapiens* from the other primates. Or, to be more precise, it is the combination of having a large brain and at the same time having a large brain mass to body mass ratio that is definitive. We can put this in perspective by noting that the quotient between brain mass and body mass for the different species normally decreases logarithmically with body size. To give a few examples: for smaller animals like squirrels the quotient is 2%, for chimpanzees 0.5%, and for gorillas 0.25%, while for larger ones like rhinoceros and oxen the quotients are under 0.1%. The exceptions to the rule are humans (with a quotient of 2.1%), dolphins (with 1.5%), and other animals which have evolved in coastal areas. The next question is: what is particular to these exceptions that has resulted in their having both large bodies and large brains? It would appear that the evolution of the brain in these animals has been able to keep pace with the evolution of a large body.

Crawford introduces chemistry as a driving force in evolution by proposing the hypothesis that the evolution of the human brain could only occur where there was an ample supply of DHA in the diet.

In essence, this means those areas near the shores of the oceans or large lakes where algae and seaweed, together with the fish and shellfish which feed on them, are readily accessible. Although controversial, this hypothesis is supported by fossil finds in Africa.

Those animals which evolved on the savanna far from the coast had to get by with the small quantity of DHA which their own livers could synthesize or with what they could get by eating other animals and their brains. Consequently, in the race to evolve a large brain predatory animals in that environment had an advantage over the herbivores. This is seen in the lion, which has more DHA than a zebra or an ox. But the difference is not sufficiently great to explain why the brain of the 'killer-ape' *Homo sapiens* is larger than the brain of the 'vegetarian-ape' chimpanzee.

As already stated, both omega-3 (DHA) and omega-6 (AA) fatty acids are required for brain formation; in the human brain they are present in approximately equal proportions. This is also true for dolphins, which in a biochemical sense are still terrestrial animals – they just happen to make their home in the sea. Hence, there is some indication that the availability of large quantities of AA and DHA is a determining factor in the growth and function of a large and complex brain.

Where do fish fit into this scenario? After all, fish have access to an enormous amount of food which is rich in DHA and their muscle fibres also contain large quantities of DHA and EPA. Nevertheless, fish have a rather small brain relative to their body size. The reason is that, in the critical phase of the development of the nervous system and brain, the fish embryo and larva has to get by with that very tiny quantity of DHA which was present in the ovum. By way of contrast, while mammals are still in the embryonic state, both AA and DHA are transferred via the placenta over the long period when the brain and the nervous system are developing.

The embryo actually draws on so many of these essential super-unsaturated fats that the brain of the mother is reduced by 3 to 5% in the last three months of the pregnancy. In addition, the newborn child has continued access to a great source of AA and DHA through the mother's milk, which is especially rich in these nutrients. If the mother is unable to produce sufficient DHA during these important stages of fetal development, it can lead to psychological problems and, in the worst cases, to blindness and reduced cognitive proficiencies in the child. This is the very cogent reason why the breast-feeding mother is advised to eat a great deal of fish or take fish oil supplements and why, in some countries, AA and DHA are added to breast milk substitutes.

Once we are adults with a fully developed nervous system and brain there is little we can do to make our brains bigger or more complex. But there is every good reason to suppose that we can maintain our nerve and brain functions at an optimal level for a longer time as we age by ensuring that we ingest sufficient DHA and AA in equal amounts, for example, by eating fresh fish or taking fish oil supplements together with vegetable oils. Research shows that individuals who eat fish regularly sharpen up their brain's efficiency and memory.

Meanwhile, there are indications that in the Western world we are well on the road to a more widespread incidence of a host of diseases involving the nervous system that cause great suffering, among them Parkinson's and Alzheimer's diseases, major depression, schizophrenia, and bipolar disorder. In addition, the growth in mental illness is alarmingly large among young people. This may well be a consequence of the Western diet, which includes far too little fish, resulting in a ratio of omega-6 fatty acid to omega-3 fatty acid that is typically a factor of twenty too high.

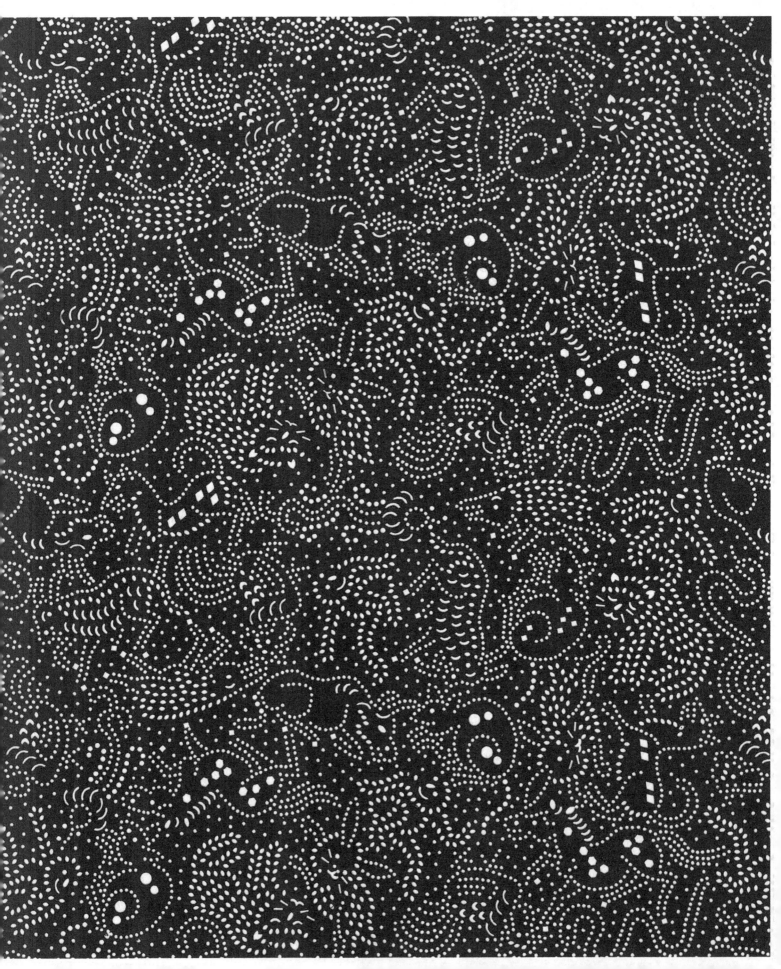

STORAGE & CONSERVATION

At the over-matured sushi
the Master
is full of regret
Matsuo Bashō (1644-1694)

FISH AND SHELLFISH

Fish and shellfish decompose very quickly after they die. For this reason they must be eaten while they are absolutely fresh or, alternatively, be deep frozen or preserved in order to increase their keeping time and conserve the healthy nutritional elements in them. One can employ a number of different methods of preservation, for example, cooking, salting, drying, smoking, marinating, and fermentation. In all cases, the taste and texture are affected.

PERISHABILITY

Fish and shellfish decompose very quickly. To a great extent, this is brought about by the often very aggressive *enzymes* of the animals, which function at low temperatures. Spoilage also happens because they are susceptible to attack by bacteria. And, finally, they contain many substances with a chemical composition that very readily undergoes a transformation, for example, unsaturated *fats* are prone to *oxidation* and *trimethylaminoxide* is quickly converted into the foul smelling *trimethylamine* ('fish odour').

Over the course of the centuries, people have made use of a number of simple methods to counteract the process of decomposition to keep fish fresh, preserve it, and prolong its keeping qualities. The most widely used methods are cooling and freezing, cooking, salting, drying, smoking, marinating, and *fermentation*. Often several of these are used simultaneously. All of them have an effect on the texture, taste, and colour of the raw product and not insignificant consequences for its nutritional value.

For the preparation of sushi, the commonest methods are cooling and freezing of fresh fish and shellfish, cooking of crustaceans and molluscs, as well as marinating/salting and fermentation of fish.

Finally, we must not forget that conservation by fermentation – packing fish in vinegared, fermented rice – is the very basis for the invention of sushi.

FREEZING OF FISH AND SHELLFISH

All living things are composed of large quantities of *water*. For example, water makes up about 60-80% of a typical fish muscle and about 80% of that of shrimp.

Freezing fish and shellfish converts a portion of the water in their tissue to small ice crystals. As ice has a greater volume than a comparable quantity of water, the consequence of this process is that *cell* walls and *membranes* can burst. When the fish is thawed, these broken cells tend to allow the fluid and dissolved substances contained in them to leak out, leading to an appreciable loss of flavour.

A collection of water molecules, H₂O, in which the hydrogen atoms (H) and the oxygen atoms (O) are shown as blue and green balls, respectively. *Hydrogen bonds* in water are shown as dotted lines between H and O.

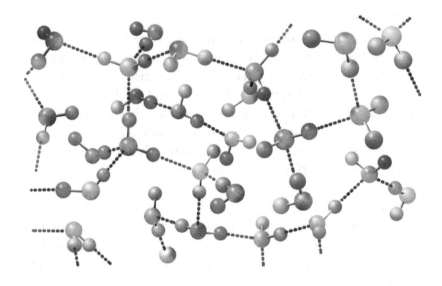

学
術
WATER plays a unique role in all biological activity. No other known substance can replace it. Water has a number of particular physical-chemical properties which are all due to the fact that the water molecule can form what are known as *hydrogen bonds*. Each water molecule can form up to four hydrogen bonds with other water molecules or with other types of molecules which have the same property. The structure and function of all *proteins* and *enzymes* are intimately connected to their ability to form hydrogen bonds both within themselves and with water.

Hydrogen bonds in water are dynamic, meaning that they form and break in order to form again later. This dynamic behaviour, in and of itself, contributes to the special properties of water. When water is frozen solid to become ice, the dynamics are weakened and the molecules are locked into position in relation to each other, often in a crystal. When liquid water in biological matter freezes to crystalline ice, the cells are destroyed and their activity stops.

It is only by employing a very special technique that flash freezes the cells at a speed of over a thousand degrees per second that the biological structure can be preserved intact. This allows cells to be studied right down to the molecular level using an electron microscope.

The cell damage done by freezing is decreased by increasing the speed at which it takes place and by taking the product to a lower temperature.

In the fresh muscle, a part of the water is bound to the *proteins*. This water is the decisive factor in determining whether the proteins can achieve the specific, folded structures that allow them to carry out their functions. Either cooling or warming sets free some of this water, causing the proteins to *denature* and to change their structure so that they can no longer function as they should. As an example, in the case of a protein that is also an *enzyme* denaturing will result in the enzyme's no longer being active. For the proteins that make up the actual muscle fibres, denaturing causes the proteins to fold in another and more compact way, drawing out some of the water. This makes the muscle fibres stiff and tough.

Oily fish do not keep as well in the freezer as lean fish, lasting only for about four months, whereas lean fish and other seafood such as shrimp can keep for up to about six months. For this reason it can be a good idea to cut away the fattiest parts when freezing an oily fish. Fish fillets which are stored too long in the freezer can develop yellow edges and surfaces, due to *oxidation* or deacidification, whereby some otherwise white substances (*flavones*) turn yellowish. The fish is then no longer suitable for consumption.

学術 GOOD ADVICE ABOUT FREEZING fish. In order to reduce damage to the surface of the fish from freezing and subsequent thawing, one can wrap the fish in tight-fitting paper or plastic film before placing it in a tightly sealed plastic bag. The plastic film ensures that ice crystals do not form on the surface of the fish, causing freezer burn. The plastic bag protects the frozen fish from frostbite and prevents it from taking on tastes from other items in the freezer. The fish should be put in the freezer carefully so that it is not crushed or constantly bumped.

One can also glaze the surface of the fish by repeatedly dipping it in cold water and each time freezing the layer of water to a thin covering of ice, which envelops it.

学術 *RIGOR MORTIS* and frozen fish. When the fish dies, oxygen is no longer transported to the tissues and no more *ATP* can be formed. *Glycogen* then becomes the only source of energy. Glycogen is gradually converted to *lactic acid* and, as the lactic acid is not able to escape, the *pH* in the muscle falls. This more acidic environment releases calcium ions (Ca^{++}), causing the muscle to seize up as if in a cramp. The resulting rigid state is 'the stiffness of death', *rigor mortis*.

Recent research has shown that it is best to freeze fish before *rigor mortis* sets in and preferably as quickly as possible after it is caught and the blood has been drained out of it. Quick freezing reduces the activity of the *enzymes*, which break down the *proteins* and *fats*. Thus, the best result is produced by the lowest temperature.

The capacity of the fish muscles to retain their liquid increases in relation to how early and how fast the freezing process takes place.

Some varieties of fish are more suitable for freezing than others. For example, the taste and texture of salmon are better preserved than those of tuna, which changes colour, loses liquid, and becomes appreciably less firm. The red protein *myoglobin* found in tuna *denatures* unless the fish is frozen to under -30°C. Consequently, the flesh of tuna that has been frozen usually has lost some of its red colour and turned brownish.

Freezing can to a certain extent help to sterilize fish that contain parasites, but bacteria are not killed by ordinary freezing. If freezing fish that has been marinated or salted, one needs to remember that the freezing point of water is lower in the presence of salt.

After its been caught and under transport, non-frozen fish stays fresh for a longer period of time if it is placed on ice (ca. 0°C) instead of stored in a refrigerator, which normally has a temperature of about 5-7°C.

When fish and shellfish are defrosted, the ice crystals melt and the water runs off. This liquid invariably contains some of the good taste substances of the fish. As a result, the thawed flesh is drier and has less taste than its fresh equivalent. How well the quality of the fish is preserved can depend on the manner in which it is defrosted. Freezing breaks down *ATP* and *glycogen*, but if any remains it can cause *rigor mortis* to set in during thawing, which in turn causes the muscles to burst. Damage caused by defrosting can be minimized by keeping the fish for a period of time at -10°C before the final thawing. This bypasses the *rigor mortis* state before the thawing is complete and minimizes the degree to which the muscles burst. All of this points to a preference for defrosting fish slowly in a refrigerator so that there is as little damage to cells and tissue as possible. Thawing it too slowly, on the other hand, can also have an unwanted side-effect because the *enzymes* present in it have more time to start decomposing the cells of the fish.

The texture of shellfish is relatively robust and is changed less by freezing than that of fish muscles. Furthermore, frozen shellfish is not damaged by being thawed in its packaging in cold water.

学術 OSMOSIS AND PRESERVATION. When salt is added to fish, the *osmotic* effect draws water out of the cells. This forms a brine around the fish that consists of water, salt, and a variety of substances which seep out of its cells. The same mechanism works on the microorganisms, for example, the bacteria found inside and on the surface of the fish. This causes water loss in the bacteria so that they dry out and, sooner or later, die. While all of this has a preservative effect, the fish itself is unfortunately affected by this treatment. A fish that has been heavily salted must first be soaked in water to make it edible.

SALTING AND MARINATING OF FISH

For sushi, salting and marinating of fish is used exclusively to kill natural parasites and bacteria in its muscles and not for purposes of conservation *per se*. Salting on its own is not normally a sufficient means of preserving fish, as there are some bacteria which thrive in high concentrations of salt. Lightly salted and marinated fish should subsequently be frozen to ensure that all parasites have been killed. This treatment will also affect the texture of the fish as the *proteins* in the muscle denature.

One can marinate fish by using acidic liquids such as vinegar, wine, or lemon juice. In some cases the fish must first be salted. Acid destroys the *cell membranes* of bacteria and other microorganisms, killing them. At the same time, the acid will inhibit fish odour from arising and impart a fresh, astringent taste. In addition, the acid softens the connective tissue in the fish muscles and reduces the length of time it takes to convert the *collagen* to *gelatine*. But there is also another and opposite effect. Marinating causes the proteins in the fish to *denature* and become more rigid, releasing some of the water. This counteracts softening. A fish that has been both salted and marinated will, therefore, often have a hard, firm surface.

Lightly marinating molluscs and cephalopods helps both to tenderize their flesh and improve their often somewhat insipid taste. If they are marinated for too long a period of time, however, they can become tough instead.

学術 *KUSAYA* IS 'ROTTEN FISH' and a great delicacy in Japan. But of course the fish is not really rotten, even if that is what the Japanese call it. Traditionally *kusaya* is made with mackerel, which is cleaned and deboned, and then placed in brine for 24 hours. Other than salt, the ingredients in the brine are usually a closely guarded secret and some claim to use a solution that is over a century old. It is most probable, however, that it contains *lactic acid bacteria*. After being removed from the brine, the fish is rinsed and put out to dry, preferably on a warm and humid summer day. It is claimed that the fish can then keep for years.

Kusaya is very reminiscent of two unique Scandinavian dishes – Swedish 'surströmming' (fermented Baltic herring) and Norwegian 'rakfisk' (fermented Atlantic salmon).

学術 'MAILLARD REACTIONS' is a term for the chemical reactions, typically associated with the browning that takes place during roasting, frying, baking, and grilling. In the course of the reactions, *sugars* bind with *amino acids*, which, after some intervening steps, results in the formation of a series of poorly characterized brown and aromatic colouring agents, called *melanoids*. These substances bring out a whole range of taste and smell sensations, from floral and herbal to meaty and earthy.

FERMENTATION OF FISH

Fermentation is a process by which microorganisms such as bacteria, convert substances in the fish, such as *sugars,* to *alcohol* or acids (for example, *lactic acid*) and produce taste substances, such as *esters* and *ketones*. It normally takes place in the absence of oxygen. In most cases, one adds salt or vinegar to the fish, just as was done when sushi was invented.

Fermentation of fish serves many purposes: to preserve the fish, to create new taste substances, and to change its texture. The actual effects of fermentation are difficult to distinguish from those of salting and marinating, in that salt and acid serve to prolong the *enzymatic* and bacterial conversion of the fish. Some of the *proteins* and *fats* in the fish are transformed to other substances with more taste and aroma. In addition, the high salt content gives rise to favourable living conditions for salt-tolerant bacteria.

HEATING OF FISH AND SHELLFISH

Only a few of the dishes in this book are cooked or fried. In general, fish and shellfish undergo the same transformations when they are heated as does meat from land animals. On the one hand, the *proteins* are *denatured* and give off water, causing the muscle to become firmer. On the other hand, the *collagen* melts to become *gelatine* and the *fats* become more fluid, which helps to make the muscles more tender and to loosen the individual fibres. The result is a fish that is easier to eat.

When fish is heated moderately, the *enzymes* in it at first produce more taste enhancing free *amino acids*. At higher temperatures, these bind to other substances, among them the *fats*, and create a series of olfactory substances that are associated with cooked fish. At even greater temperatures, which are achieved by roasting or frying, the amino acids and *sugars* create the chemical compounds that are characteristic of browning of food. The chemical reactions which result in browning are called *Maillard reactions*.

Blanching with boiling water is used to tenderize fish skin which has small scales or none at all. It also serves to impart a slightly sweeter taste, because the process releases free *amino acids*.

When crustaceans and molluscs are heated, a range of diverse, characteristic taste and aroma substances are brought forth. These are quite different from those associated with heated fish.

When crustaceans are cooked, substances are formed which are compounds of *sugars* and *amino acids* and which resemble those that form on the surface of meat and fish when they are browned. In addition, short-lived chemical compounds, for example, *bromophenols,* which we associate with the smell of the sea, are released.

The taste of crustaceans is more intense when they are cooked in their shells, because the shells also contain taste substances. In addition, the shells encase the *proteins* and *sugars* in the muscle meat so that they do not leach out into the cooking water.

Molluscs such as mussels lose some of their delicate, sweet taste when they are cooked because the otherwise tasty free amino acids become bound. On the other hand, heating releases other intense aromatic substances, especially *dimethyl sulphide*, the smell of which is familiar from warm milk.

WHY DO COOKED CRUSTACEANS TURN RED?

The tails of crustaceans such as shrimp, lobster, and crayfish consist of a large, powerful, transversely striped muscle which is white. It is built up of bundles of fast muscles, which are activated when the animal needs to move. Their shells are actually their skeletons, which, in contrast to those of bony fish, are found on the outside.

The shells of crustaceans are blue-green, due to a chemical complex, *crustacyanin*, which consists of a *protein* to which is bound a reddish orange pigment, *astaxanthin*, known from the red colour of salmon. As long as the astaxanthin is bound to the crustacyanin complex, no red colour is visible. When the crustacean is cooked, the crustacyanin is *denatured*, releasing the astaxanthin and bringing out the red colour. The part of the muscle meat located under the shell also contains some astaxanthin and also turns red.

学術 A GOLDEN RULE. Fresh fish must be heated (cooked/steamed/fried) for only a very short period of time and preferably not at all! But if you are in doubt as to whether the fish contains bacteria, you must heat it to over 85°C in order to be sure that all the microorganisms have been killed.

TSUKEMONO – THE ART OF PICKLING

The term *tsukemono* includes a wide range of wholly or partly preserved products made from vegetables and fruits, employing techniques of salting, souring, or fermentation. We would call them pickles. The pickling produces new flavours and preserves the nutritional value. *Tsukemono* are eaten at virtually every Japanese meal. Pickled ginger (*gari*), an important condiment with sushi and *sashimi*, is a typical example of *tsukemono*.

THE JAPANESE WAY OF PICKLING

Tsukemono is much more than a method of preserving food. It is just as much a means of preparing raw foods – first of all, so that they become edible and keep their nutritional value, and secondly, to make them appealing by imbuing them with beautiful colours and pleasant, exciting aromas. In addition, *tsukemono* introduce taste sensations that are complementary to the other dishes with which they are served.

Many types of *tsukemono* are not intended for long-term storage. Hence, one can use smaller quantities of salt, *vinegar*, and *alcohol* and possibly a shorter *fermentation* period. But there are also *tsukemono* which can be kept for several years.

Ideally, conservation of a vegetable or fruit takes place in three stages. First, the *cells* of the plant are broken apart, next its own *enzymes* must be made harmless, and, lastly, an environment in which microorganisms cannot thrive must be created. The Japanese art of pickling in brine, *tsukemono*, comes close to this ideal.

Pickling sets in motion a *fermentation process* that takes place in a saline environment using *lactic acid bacteria*. Heat is not needed and, in principle, pickling can take place throughout the year, although with varying results. In order to create the acidic conditions in which the lactic acid bacteria thrive, it is necessary to use an acid such as *vinegar*. Fermentation results in even more acid, which results in more thorough pickling.

What is special about pickling is that the bacteria which cause the fermentation are found naturally on the plants and the process starts as soon as the plant cells have been destroyed by the brine. Fermentation converts the *sugars* in the plants while leaving the rest of their nutritional elements intact. In particular, *vitamin C* is preserved and is protected by the carbon dioxide, which is a by-product of fermentation. The pickling process is so intense that other microorganisms have no chance to get in on the action.

Modern Japanese pickling jar for the preparation of *tsukemono*. The jar is equipped with a plunger which can press the contents together so that the brine seeps out and covers them.

珍
談
IT IS SAID THAT the gentle and careful way that Japanese handle vegetables and salt in preparing *tsukemono* has its roots in the *Shinto* concept that vegetables and salt are gifts given by the gods to humans in order to ensure their good health.

学
術
NUKA-ZUKE AND *MISO-ZUKE* are traditional methods used in the Japanese way of pickling, *tsuke-mono*. In these cases, the process is driven by an additive that actively causes *fermentation*, for example, mash left over from the production of *sake*, *miso*, or soy sauce, or more directly by introducing fresh rice bran (*nuka*) or *miso*.

学術 THE TEXTURE OF *TSUKEMONO* made from vegetables, for example, cucumbers and eggplants, is crisp and crunchy when they turn out well. Crispness can be enhanced by using sea salt, as it also contains calcium and magnesium salts. The Ca^{++} and Mg^{++} ions from these salts form bonds between the carbohydrates on the cell walls, which in plants are covered with a *polysaccharide* called *pectin*. Pectin is the substance that helps to set fruit jelly and marmalade naturally.

学術 BE CAREFUL if you pickle using vinegar without allowing *fermentation* or using enough salt. In this case, the pickled products last for only a short while and they must be stored under refrigeration.

For pickling, it is best to have a crock with a plunger-style movable lid which can be pushed down to compress the contents. The liquid which seeps out forms a brine that covers the vegetables or fruits. This limits contact with oxygen in the surrounding air, which would foster the growth of unwanted bacteria. Which type of fermentation microorganisms will dominate the pickling process depends on the salt concentration and the temperature.

The Japanese pickling tradition encompasses not only cucumbers and other vegetables, but also stone fruits such as plums and Japanese apricots, which are made into *umeboshi*.

In Japan one can get an untold variety of *tsukemono*, but in other parts of the world their availability is limited unless one makes them at home. The products most commonly sold are pickled ginger, cucumbers, eggplants, radishes, and plums.

Tsukemono and health are nutritionally linked, because *tsukemono* have a significant fibre content, are rich in vitamins (especially vitamin C, but also B_1 and B_2 in *nuka-zuke*), and contain large amounts of calcium. In addition, many of the natural *enzymes* of the plants are preserved because the ingredients are not heated at any point. These enzymes act in concert with the acid and salt content in *tsukemono* to whet the appetite and improve digestion.

But there is also a negative aspect – eating too much salt is inadvisable and can lead to high blood pressure. One way to counteract this is by substituting *rice vinegar* for a certain proportion of the salt or by adding *alcohol*.

Another term for *tsukemono* is *oshinko* which literally means 'new fragrance'. It indicates that Japanese pickling is no longer as much a way to preserve food, as it is a means of compensating for the relative paucity of ingredients by transforming them into several new incarnations and then regarding them as separate entities. There is a *Zen* aspect to this – it is a way of maximalizing minimalism.

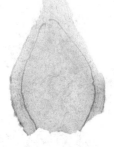

Thin slice of pickled ginger (*gari*).

学術 PICKLED GINGER from Thailand often has a sharper taste than Japanese *gari* and it is often more coarsely cut. I find it less suitable for sushi, but perfectly good in warm dishes.

珍談 IT IS SAID THAT Japanese housewives were once called *nuka-miso*, which could probably be translated as 'smelly women' because their hands took on an odour from the daily stirring of the fermentation medium, *nuka-doko*, used in the preparation of *tsukemono*.

PICKLED GINGER (*GARI*)

Gari is probably the type of *tsukemono* that is best known in the Western world. It is made from fresh ginger root (*shōga*) and is an important accompaniment every time sushi and *sashimi* are served. The pickled ginger has a very fresh, sharp, and slightly citrus-like, but also sweetish, taste, which serves to clean the palate in between bites. This helps one to discern the often subtle differences in the actual taste of the various fish and shellfish, as well as those caused by the methods of preparation. The sharp taste of ginger is due to *gingerol*, which is in the same chemical family as *piperin* and *capsaicin*, the taste substances that give black pepper and chili their strong spicy flavours.

Gari is prepared by peeling and thinly slicing fresh ginger root, which is then pickled in salt, rice vinegar, and sugar. It is naturally a pale yellowish colour, but the pickled product is often coloured using red *shiso* to give it a rosy pink appearance. It is stored in its brine, which can also be used as a dressing for fresh cucumbers and avocados or when marinating salmon.

Adding *gari* is a superb way to enhance the taste of a large number of side dishes; for example, it can be combined with avocado, fresh cucumbers, zucchini, and salmon.

PICKLED RADISHES (*TAKUAN-ZUKE*)

The large white radish, *daikon*, also known as Chinese or Oriental radish, is used to make a *tsukemono* called *takuan-zuke*, a favourite among the Japanese. It is an example of *nuka-zuke*, in which rice bran is introduced as the *fermentation medium, nuka-doko*.

Nuka-doko (*toku*) is a mash containing rice bran (*nuka*) and possibly *miso*, into which water, salt, kelp (*konbu*), sugar, and sometimes red chili peppers and orange or apple peel are mixed. *Fermentation* is initiated by adding carrot peelings, cabbage, or the green tops of the radishes. Once this has been established as a starter culture, it can be used to make *tsukemono* by immersing the vegetables or fruit in the mixture. There are many variations on this recipe, and each Japanese person who makes *tsukemono* is said to have his or her own secrets. *Nuka-doko* can be used over and over and keeps for a long time provided that it is looked after properly and stirred well every day.

The production of *takuan-zuke* requires that the radishes first should be hung up to dry and then salted. This causes the radishes, which are typically 30-40 cm long, to become pliable so that they can be bent and packed tightly in a wooden barrel together with their cut-off green tops. The fermentation medium is placed around them.

A stone is placed on top of the movable lid of the barrel in order to press the contents together. Now the *fermentation* process sets in. After a few weeks, so much brine will have been formed that it covers the radishes completely. The radishes are then left in the brine for a couple of months before they are ready for use. Before serving, the brine and the remnants of the *nuka* are rinsed off.

When completely fermented, *takuan-zuke* take on a pale yellow or yellowish brown colour and normally the pieces are sliced thinly before being served. Sometimes yellow food colour is added to commercially prepared versions.

Takuan-zuke are good with warm rice, in soup, or simply as a side-dish together with other types of *tsukemono* at a sushi meal. They are also used in a popular form of *maki*-zushi, *oshinko-maki*.

PICKLED CUCUMBERS AND EGGPLANTS

Cucumbers and eggplants are some of the most commonly used vegetables for making *tsukemono* in brine, known as *shio-zuke (na-zuke)*. *Shio* means salt from the sea, which originally was the only source of salt as there are no underground salt deposits in Japan. This type of *tsukemono* is often used to cleanse the palate between the courses of a meal.

I am particularly fond of *tsukemono* made from cucumbers. They acquire an interesting texture, which is crunchy on the outside but still elastic and soft inside. Unripe cucumbers are used and they are placed in a 4-6% salt solution (*pH* 3.4-3.8) or directly in salt. They are then fermented in a crock, to which sugar may be added, at a temperature of 18-20°C. *Lactic acid*, carbon dioxide, *alcohol* (*ethanol*) and a variety of aromatic substances are produced. When finished, crushed sesame seeds may be added to the cucumbers. Small Japanese cucumbers, *kyūri*, are particularly suitable for making *tsukemono* because they have small seeds and are very crisp.

Takuan-zuke is a type of *tsukemono* made from radishes (*daikon*).

珍談 IT IS SAID THAT *takuan-zuke* were invented in the 17th Century by Takuan Soho, a respected and influential *Zen* Buddhist monk, and that they bear his name.

珍談 IT IS SAID THAT the custom of using *tsukemono* as a side-dish at meals goes back to the 10th Century in Japan, when the noble families held contests to guess the smell of various types of incense. In order to sharpen their sense of smell while contests were in progress, they ate *tsukemono*, also called *kōnomono*, which means 'a thing in connection with incense'.

Dried, sometimes lightly toasted or salted, red *shiso* leaves are often added to cucumber *tsukemono*, resulting in an accompanying dish with a beautiful colour.

Shiba-zuke is a type of sour and salty *tsukemono*, consisting of eggplants or a mixture of cucumbers, eggplants, and radishes, which have taken on their colour from red *shiso*. I have had good results using green tomatoes as an ingredient in *shiba-zuke*.

Small eggplants (*nasu*) also make excellent *nuka-zuke*. They are fermented in the same manner as radishes.

調理法 PICKLED GREEN *SHISO* LEAVES. 100 grams (about ¼ pound) green *shiso* leaves, 1 litre (4 cups) water, and 30 millilitres (2 tablespoons) salt. The *shiso* leaves are picked with their stems on, washed thoroughly, and made into bundles of ten leaves tied with a cotton string around the stalks. The leaf bundles are placed in a pickling jar with a lid that can be pushed down. Salt is mixed with the water and poured over the leaves, which are then pressed down with the lid and left for about 24 hours. The salt water is drained off and the leaf bundles are pressed lightly to remove excess moisture. The leaf bundles are then placed in a container with a lid that can compress them. A little salt is sprinkled in the bottom and the bunches are placed in neat layers with salt in between and the air squeezed out. The salted green leaves keep in the refrigerator for up to a year. When the green *shiso* leaves are to be used, they are soaked in cold water and then dried carefully by placing them on pieces of paper toweling.

Salted red and green *shiso* leaves

The selection of fresh herbs found in Western fruit and vegetable shops and markets rarely includes *shiso* plants, but they are relatively easy to grow as they are members of the mint family. The trick is finding the seeds or the plants.

By pickling harvested *shiso* leaves in brine, it is possible to keep them well past the growing season. In addition, salted red *shiso* leaves are useful for the salt preservation of plums, *umeboshi*.

Whole salted green *shiso* leaves can be folded around a rice ball or chopped and incorporated into rice balls or sprinkled on top of them. They can also be combined with other pickled vegetables.

Salted red *shiso* leaves can be chopped and used to make *umeboshi* or served as *yukari*, for example, sprinkled over warm rice or oven dried eggplants.

調理法 **Salted red *shiso* leaves.** 100 grams (about ¼ pound) red *shiso* leaves, 15 millilitres (1 tablespoon) salt.

Shiso leaves are picked with the stem on and washed thoroughly, then put in bundles of ten leaves tied with a cotton string around the stem. The bundled leaves are placed in a pickling jar that has a lid with an adjustable pressure plate. Salt is sprinkled on the leaves and sufficient water is added to cover the leaves. The pressure plate is adjusted to hold the leaves down in the water and everything is left for 3-4 days. Then the brine is drained from the leaf bundles, which are compressed further to remove the dark fluid that seeps out of them. The leaf bundles are transferred back to the cleaned pickling jar and covered with plum vinegar (derived from making *umeboshi*) or white wine vinegar to which a little salt has been added. Allow the leaves to remain pressed together for a couple of days. Store the leaf bundles in the refrigerator. Some of the red colour of the leaves is lost in the pickling process.

調
理
法

Umeboshi – pickled plums. This recipe has been elaborated to suit conditions where I live. I have had a great deal of pleasure in experimenting my way forward with the help of my good physics colleague, Per Lyngs Hansen, and 'the plum ladies' who sell their fruit on the market square in Odense, my home town.

Wash 1 kilogram (about 2 pounds) of Italian prune plums thoroughly and place them on a bamboo sieve to dry for three days. The plums are then put in a pickling jar with a lid that has an adjustable pressure plate. It is important that the process is carried out under sterile conditions and that the jar is scalded before use. Sprinkle about 100-150 grams (½ cup) of salt on the plums and leave them to rest for two days, but turn them over gently a few times. At this point pressure is applied for 4-5 days.

Ten red *shiso* leaves are then added to provide colour and flavour and to serve as a preservative. If dried or salted *shiso* leaves are not available, *yukari* that is a mixture of crumbled toasted red *shiso* and salt can be substituted. About 30 millilitres (2 tablespoons) of *yukari* are used for this quantity of plums. The plums are now allowed to continue pickling for a week. True *umeboshi* are then removed from the brine and dried, but I prefer to leave mine in the pickling juices and store them in the refrigerator.

Pickled Japanese apricots or plums (*umeboshi*)

Umeboshi are traditionally made from *ume*, which are not really true plums, but rather a type of apricot (*Prunus mume*) grown in Japan. As this fruit is not suitable for eating fresh, it should be dried and pickled in brine, deriving colour and flavour from the red *shiso* used in the process. The *perilla aldehyde* found in red *shiso* also acts as such an effective preservation agent that properly made *umeboshi* can easily keep for up to a year. *Umeboshi* are served with the small pieces of red *shiso* which have stuck to their surface.

The taste of the *umeboshi* is unmistakably fresh and appetizing. They are eaten whole or cut in smaller pieces and placed on warm rice or inside *maki*-zushi. As they are usually rather salty, a whole one, or even a part of one, is plenty. In dried and crumbled form, *umeboshi* are also sprinkled on warm rice. The sour brine in which the fruit is pickled can be reused to pickle other produce, such as radishes, eggplants, or cucumbers.

Japanese *ume* are not readily available outside of Japan and the closest alternative is probably regular plums. Not all types are appropriate for pickling in brine. After some experimentation, I have found that the common European plum (Italian prune plum or *Prunica domestica*), which is quite firm, is most suitable. It is important that the plums do not have a skin which is too thin or delicate, but is sturdy enough to hold up during pickling or storage. Greengage plums can also be used, but they seem to keep less well and the skin has a tendency to break.

Many stories are linked to *umeboshi* as the key to a healthier life. Evidence of this is that many Japanese eat a single dried, salted *umeboshi* as part of their breakfast. A traditional Japanese meal is often concluded by serving a small bowl of warm rice with *umeboshi*. It is refreshing and thought to enhance digestion. It is said that eating one *umeboshi* plum every day at breakfast keeps the digestive system in good order. The *samurai* followed this practice so that they would always be on the alert and ready to fight. The physiological explanation for this could be that *citric acid* in the fruit helps to convert lactic acid in the stomach to carbon dioxide and water. This prevents tiredness and stimulates the digestive tract. In addition, the citric acid increases the uptake of calcium in the intestines.

TOOLS, PREPARATION & PRESENTATION

*Pressing sushi
after a while
a lonely feeling*
Matsuo Bashō (1644-1694)

TOOLS FOR MAKING SUSHI

Most of the equipment one needs for making sushi can be found in the average modern kitchen. Only a very few specialized tools are needed; these include a sharp sushi knife, a bamboo rolling mat, and an electric rice cooker. On the other hand, it can certainly enhance the pleasure and the aesthetic experience inherent in preparing sushi if one acquires some Japanese kitchenware and a few authentic tools.

OLD AND NEW MAKE FOR GOOD *WABI SABI*

Most of the kitchenware and tools that one uses to prepare sushi are simple and produced from natural materials, such as bamboo and wood, which absorb excess moisture, are hygienic, and do least harm to raw ingredients. Wood or bamboo sieves are used to allow excess liquid to drain from vegetables and seaweed that have been washed or soaked in water. Cooked rice is cooled in wooden tubs and vegetables and fish are cut up on wooden boards. The principal tool of the sushi chef is a special type of knife, whose origins can be traced right back to the legendary sword of the *samurai*. Platters and bowls used to serve sushi and cups for drinking tea are inspired by centuries of *raku* tradition.

Besides these are all the modern and practical pieces of equipment, among them items made from synthetic materials and automated appliances which can function on their own.

The Japanese kitchen is a reflection of Japanese society – a mixture of older items that are bound up with venerable traditions and rituals and hi-tech appliances such as rice cookers and microwaves. What is distinctive is that the Japanese, in this regard as in the broader context of their way of life, manage without difficulty to make the various elements function together harmoniously despite the many centuries that separate their points of origin. When the electric rice cooker was introduced it was said that it sounded the death knell for Japanese culture. But today there are few Japanese homes, restaurants, or sushi bars that do not use one.

It is good *wabi sabi* to combine the old and the new, with respect for the old and worn and with an eye for manufacturing new articles in accordance with traditional design principles concerning simplicity and beauty.

KNIVES (*HŌCHŌ*)

Sharp knives with the right thickness, length, and shape of handle are the most important tools possessed by a sushi chef. Every serious sushi chef has a personal set of knives which he brings with him and which he will only grudgingly allow others to use. More than twenty different types of knives are found in a professional Japanese kitchen. For the preparation of sushi, however, one needs an assortment of three at most.

The sushi chef's knives: a sturdy knife with a curved cutting edge for sectioning fish and cutting through bone (*deba-bōchō*); a heavy knife with a wide blade and an straight edge for slicing and peeling vegetables (*usuba-bōchō*); and a long knife with a narrow blade and an almost straight edge for slicing fish, *maki* rolls, and making decorative garnishes (*yanagiba-bōchō*). The knife handles are usually made from untreated wood and they are fitted to the blade by a bone cuff.

The *yanagiba-bōchō* shown above has a pointed end that is characteristic of Osaka and the western part of Japan. In Tokyo and the eastern part of the country, the traditional sushi knife has a blunt end (*tako-biki*).

Tako-biki, a sushi knife with a blunt end.

The top side of the blade of a classical fish knife for sushi and *sashimi* (*yanagiba-bōchō*), showing the imprint of the cutler (Kiya in Tokyo). The wavy flame-like pattern of this forge-welded knife is seen clearly.

If one buys fresh fish that has been sliced and deboned, one can actually get by with just the classical fish knife (*yanagiba-bōchō*), which is a long knife with a narrow blade and an almost straight cutting edge.

Sushi and *sashimi* knives have only one cutting side, which is always the right hand one. Knives sharpened in this way cut more quickly and cleanly than knives with two cutting sides.

The heavier the knife and the longer the blade, the easier it is to cut pieces of uniform thickness. The knife must be as sharp as possible, so that it can cut with least difficulty.

There are three ways of cutting:

1. The knife is drawn toward the body and the weight of the knife performs the cutting operation. This is how one slices fish fillets for sushi or *sashimi*. The entire length of the blade is utilized.

2. The knife is pushed away from the body and the smooth cutting edge of the knife is held vertically in relation to the item being cut. This is how one cuts the pieces of a *nori* leaf or a *maki* roll.

3. The knife is pushed forward, applying slight pressure. This is how one slices vegetables, for example, cucumbers.

Good cutting style involves holding the knife close to the cutting board and working into a good rhythm.

Proper sushi and *sashimi* knives are made from carbon tempered steel. Steel is composed of iron and a number of other metals. The carbon content makes the knife hard and consequently sharp, but also brittle. If one drops a knife of this type on a hard table or floor, there is a real risk of snapping off the point or of making dents in the cutting edge.

I have a magnificent Japanese fish knife made from steel that is called 'folded' or 'forge welded'. This type of steel is recognizable from the characteristic wavy flame-like appearance of the surface of the blade, which is also beautifully patterned and resembles a classical

Damascus knife. These knives are very strong and have a special hardness that allows them to be sharpened to a high degree.

It goes without saying that a real sushi knife is washed by hand, in clean water without soap. Never use a dishwasher. It is important that the knife should be thoroughly cleaned after use and that no traces of fats remain on the blade or the handle. It must be dried completely with a cloth, as moisture will cause a normal carbon tempered steel knife to rust. The knife can be stored in a knife block, but care must be taken so that the cutting edge does not touch the block as this could dull it. By preference, sushi knives should be stored in a wood or cardboard knife box or a special wooden sheath made for this purpose.

A sushi knife has to be kept sharp and it is a good routine to sharpen it at regular intervals using an appropriate whetting stone (*toishi*). It is inadvisable to use a sharpening steel or a sharpener with rotating wheels as it can be difficult to keep the knife as a constant angle. This could cause the steel and the cutting edge to become uneven and, in the worst cases, might result in nicks. It is best to use a flat, hard whetting stone made of carborundum (silicon carbide) or one that is diamond coated. A very dull knife is first sharpened on a coarser emory stone and then on one or more whetting stones which are progressively finer. One should strive to sharpen the entire length of the knife blade in one stroke.

Sushi knives are sharpened on one side only, the cutting side, which is the right hand side. The small burrs that form on the other side are removed at the end by drawing the flat side of the knife lightly over the stone, holding its surface flat against it.

Of course one can also use a good quality knife made from stainless steel. Stainless steel knives are easy to maintain. They are difficult to sharpen, but their edge lasts a longer time than that on carbon tempered steel.

A good carbon tempered sushi knife or a forge welded knife is an expensive proposition, but it is worth the money if one values a superior tool. A high quality knife lasts a lifetime and one will derive pleasure from it every time it is taken out of its box. Note that a good whetting stone (*toishi*) does not come cheap either.

Deba-bōchō with a tightly fitting wooden sheath. During transportation, the knife is secured in the sheath by plugging in the wooden peg.

学術 FORGE WELDED STEEL is produced by a labour-intensive process. Different iron-containing metallic mixtures with contrasting compositions are joined by repeatedly folding them together, heating them, and applying pressure. Polishing and acid etching bring out the flame-like pattern, adding a unique visual dimension.

Forge welded steel is often incorrectly described as Damascus steel. The latter is made by another process for which, as far as is known, there is no detailed description and there are no longer any practitioners who know its secrets. At present, many attempts are being made to rediscover this ancient technique. Damascus steel is associated with the magnificent swords of the Saracens which came to be known in Europe during the time of the Crusades. These swords were prized for their strength, sharpness, and beautiful, flame-patterned blades.

Toishi, a fine whetting stone for sharpening sushi knives.

A simple electric rice cooker (*suihanki*) without electronic programming and only two functions: 'cook' and 'keep warm'. It automatically switches to 'keep warm' when the moisture content in the rice indicates that it is ready.

学
術 WOOD OR PLASTIC? Many people are of the opinion that a cutting board made out of plastic is more hygienic than a wooden one. This is not actually correct. Microorganisms apparently thrive better in the cracks of a cutting board made of synthetic material. Wood is more hygienic due principally to the fact that the bacteria lodged in its cracks have worse living conditions than on plastic because wood draws the water out of them.

RICE COOKER (*SUIHANKI*)

When I first started to make sushi, I cooked rice in an ordinary saucepan. It can be done, but it is difficult and one has to be constantly vigilant so that the rice does not burn or overcook.

An electric rice cooker (*suihanki*) is the solution. I prefer a simple version without too many electronic features and a whole bunch of programs for making all sorts of different rice dishes. All I want is one that has a 'cook' and a 'keep warm' button. A rice cooker has a tight fitting lid, which means that the cooking is carried out under slightly increased pressure, decreasing the cooking time.

The rice cooker shuts itself off when the moisture content in the rice indicates that it is ready. But if one has not stirred the rice a few times while it was cooking, one cannot be sure that it is ready to make sushi when the cooker shuts off. Experience has taught me that the rice is just right for sushi if one lets it stand on 'keep warm' for a couple of minutes after it has shut off. At this point, the rice has to be taken out of the cooker so that it does not end up overcooked.

CUTTING BOARD (*MANAITA*)

All Japanese cuisine is predicated on the use of a cutting board (*manaita*), which is where most of the action takes place. This is reflected in the Japanese language as the word for chef, *itamae*, literally means 'in front of the board'.

I prefer to use a large, massive wooden board which is quite hard. It must not be used for raw ingredients with a strong smell, such as onions, as the odour can penetrate the board.

The wooden cutting board should be soaked thoroughly in water and wiped with a damp cloth before use. Moisture in the wood prevents things from sticking to it and the damp surface makes the wood softer, which helps to prevent the knives from becoming dull too quickly. The cutting board must be cleaned carefully after use in clean water and preferably without using very much soap.

WOODEN TUB AND WOODEN PADDLE (*HANGIRI* AND *SHAMOJI*)

Apart from my knife, rice cooker, and rolling mat, the wooden tub (*hangiri*) for cooling cooked sushi rice is probably the item I would least like to do without in my kitchen. The wooden tub has low sides and is made like a barrel with staves that are held together with copper hoops. Tubs of this type can be had in different sizes. An appropriate size for a normal household is one with a diameter of ca. 35 cm. The tub is made from cypress wood, which has a pleasant smell when it gets damp.

For cooling, newly cooked sushi rice is placed in the wooden tub, in which it is then mixed with salt, sugar, and rice vinegar, a subject to which we will return later.

The tub is soaked in water before use and then dried with a damp cloth. This is done so that the rice grains will not stick to the wood. After use, the tub is washed in clean water without the use of soap, and then it must be dried thoroughly. In order to prevent the staves of the tub from drying out, shrinking and then falling apart, it is necessary to use it regularly.

The classical, Japanese wooden paddle, *shamoji*, takes on an important role, at once practical and symbolic, in the Japanese kitchen. The paddle is soaked in water before use so that the grains of rice will not stick to the dry wood. It is used to turn the sushi rice in the wooden tub when it is cooling and to stir in the rice vinegar. In addition, it used for serving warm rice at the table. One can also use the paddle to force cooked vegetables through a sieve to make puree. Of course, a plastic spatula is just as good for these purposes.

Shamoji is the classical Japanese symbol for the status and authority of the housewife in running the home. When a younger woman assumes responsibility for a household, the *shamoji* is handed over to her by the older housewife, who is now relieved of this responsibility. It is a sign of recognition that the younger woman is deemed worthy of taking over the task.

The traditional low-sided wooden tub, *hangiri*, used for cooling freshly cooked sushi rice.

学術 *O-HITSU* is another traditional Japanese wooden container which is used to store and serve warm rice. The container is closed with a tight-fitting wooden lid. The modern rice cooker has to a large extent overtaken the role of this wooden vessel.

Wooden paddle, *shamoji*, for stirring and serving cooked rice. This paddle is made of bamboo.

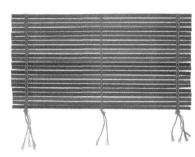

Bamboo rolling mats (*makisu*) for making *maki*-zushi. The long mat is used to prepare thick *futomaki* rolls and the short one for thin *hosomaki* rolls. The smaller mat is shown with the flat outer side of the bamboo strips on top.

BAMBOO ROLLING MAT (*MAKISU*)

A rolling mat (*makisu* or *sudare*) is absolutely essential for making *maki* rolls with *nori* leaves. The mats used to prepare *maki*-zushi come in two sizes, a square one about 25 cm × 25 cm for making the thick rolls (*futomaki*) and one half the size, 12.5 cm × 25 cm, for the thin *hosomaki* rolls. I myself use a large *makisu* for preparing all types of *maki* rolls.

The rolling mats are produced from thin strips of bamboo that have a flat, matte green side, which is the outer surface of the bamboo stem, and a rounded inner side. The strips are woven tightly together with cotton strings knotted at one end of the mat. On some rolls, the bamboo strips are completely cylindrical and these are used to shape omelettes (*tamago*), leaving a fine lined pattern imprinted on them.

The mat must be cleaned very carefully after use, with a stiff brush, so that no food bits remain in the narrow spaces between the bamboo strips and in the strings that hold it together. It must also be dried completely once it is clean to prevent its being attacked by fungi.

A DAMP, CLEAN CLOTH (*FUKIN*)

Although it sounds like a triviality, one needs a damp, clean cloth (*fukin*) when making sushi. It is, however, used very differently from an ordinary kitchen dishcloth.

One needs to have the cloth at hand at all times as it serves a number of functions. First, and foremost, it is used to wipe the cutting board, knife, and other tools. It is also used both to clean and to dampen one's fingers. In addition, the cooked sushi rice in the wooden tub is covered with a damp cloth so that the top layer does not dry out.

The cloth must always be completely clean and, therefore, it must be rinsed frequently in cold, clean water and wrung out. I prefer to use white terry towel washcloths when I make sushi and usually have several in use at the same time.

Bamboo sieve (ZARU)

Sieves woven from bamboo (*zaru*) have a number of uses when one is making sushi. The sieves are cheap and come in several sizes. Some are round, others square, and there are both shallow and deep sieves.

Round-bottomed, shallow bamboo sieve (*zaru*) is used in numerous ways in the kitchen, for example, to rinse rice.

A fine meshed sieve is ideal for rinsing rice before it is put in the rice cooker. It can also be used to drain water from salad leaves and vegetables. A flat round-bottomed sieve fulfills an important function when one is tenderizing fish skin. Boiling water is poured over the fish, which is covered with a cloth. Because one does not want the process to affect the flesh of the fish, it is important to use a *zaru* so that the boiling water can quickly drain away.

I also use my sieves when I am air drying *shiso* leaves.

Tweezers (HONE NUKI)

Some types of fish have many small bones which must be extracted from the fillet before it can be sliced for sushi or *sashimi*. A pair of tweezers (*hone nuki*) with a wide flat tip is handy for this purpose as it permits one to get a good grip on the bones. Fresh water fish, in particular, can have many small bones. And in some fish, for example, mackerel, the bones are very firmly attached where the connective tissue between the bones and the skin is especially strong.

Tweezers (*hone nuki*) are handy for extracting small bones from a fish fillet.

Grater (OROSHI-GANE)

A grater (*oroshi-gane*) is essential for shredding ingredients such as radishes or *wasabi*, should one be so lucky as to find a genuine *wasabi* plant. Graters are made from metal, plastic, and porcelain, and range from coarse shredders to ones so fine that they reduce the raw ingredients almost to a mush.

Bamboo skewers (KUSHI)

Thin bamboo skewers (*kushi*) are available in different lengths and thicknesses. They are used in connection with sushi to help food keep its shape. For instance, large shrimp are skewered to prevent them from curling up when they are cooked.

Bamboo skewers are also commonly used for grilling pieces of meat.

Traditional Japanese grater (*oroshi-gane*) made of tinned copper is used for fine shredding of ingredients such as *wasabi*. A more primitive type of *wasabi* grater is made from a piece of shark skin glued onto a wooden board.

Three part wooden mold (*oshibako*) for preparation of pressed sushi, *oshi*-zushi. The mold consists of a footed bottom, a frame, and a lid which fits inside the frame and can be used as a pressure plate. The lid has two cross pieces that are useful when one is pressing down and on which one could place a weight.

Hand mill for crushing roasted sesame seeds.

Omelette pan (*tamago-yaki-nabe*) for making omelettes (*tamago-yaki*).

MOLD FOR *OSHI*-ZUSHI (*OSHIBAKO*)

When one makes pressed sushi (*oshi*-zushi), it is necessary to have a mold which can both hold the ingredients together and which can easily be taken apart so that the pressed sushi can readily be removed without losing its shape and becoming ragged around the edges.

The *oshibako* is a special three-part wooden mold. It is a rectangular box with a bottom that can easily be removed and which has a lid that acts as a pressure plate to squeeze the contents together. Typically, the mold is carefully crafted from cypress wood and comes in a variety of sizes. The standard size has inner dimensions of ca. 8 cm × 17 cm so that one can easily fit a suitably trimmed half sheet of *nori* into it.

Before use the mold is soaked in water and then dried thoroughly with a cloth. The moisture remaining in the wood ensures that the rice does not stick to it.

SESAME MILL

Sesame seeds must be crushed in order to release the maximum flavour and aroma found in the sesame oil, but this should be done only just before use to derive the greatest benefit from these substances.

For this purpose one can use an ordinary mortar and pestle or a small hand mill. Small electric mills are also on the market and they have the advantage that they can be used with one hand.

OMELETTE PAN (*TAMAGO-YAKI-NABE*)

It is an advantage to use a rectangular pan (*tamago-yaki-nabe*) for making omelettes (*tamago-yaki*) to be used for sushi because the cooked, folded omelette will have even sides and be of a more uniform thickness. Of course, one can use a regular round pan, but then it will be necessary to trim the omelette after it is folded over.

How does one become a sushi chef?

On 'stealing the art' from the master

Can one become a sushi chef if one is not Japanese? Yes, of course, but most sushi chefs, especially the really outstanding ones, are Japanese. It does not lie in the genes, but in the culture, just in the same way as Japanese can become good ice hockey players, but the best ice hockey players are North American or European. Similarly, and according to tradition, it is only men who can become sushi chefs.

It is a little misleading to describe a person who prepares sushi as a cook, as there is little cooking in a sushi bar. A master chef in a sushi bar is referred to as an *itamae* which means 'he who stands in front of the board', that is to say, the table where the sushi is prepared. But often a sushi chef is called a *sushiya*, an expression which also encompasses the sushi bar itself. By established custom, a *sushiya* has less prestige than a real Japanese chef, *itamae*, because the former uses only two or three knives for his work, whereas the latter uses ten different ones. An assortment of knives is the hallmark of the chef.

The classical training of a true *sushiya* or *itamae* follows the Japanese tradition for all the arts. It is said that to become a good *sushiya* one must start as a teenager in order for the art to become the very essence of one's being, a second backbone. One learns according to the principle of *nusumu no gei*, namely, to learn by observing the master and 'stealing the art', not by instruction.

Initially, the apprentice is very much in the background. For the first two or three years, while observing the master performing his art, the apprentice also learns to gut and scale fish, to clean up, and to be of assistance. There is no doubt that in former times the young trainee was seriously exploited during this period of initiation, under the pretence of teaching him respect for people and materials.

In the succeeding period, the apprentice is allowed to cook rice and practice shaping it. If the young aspirant has by this time indicated that he has both talent and patience, he is allocated a far corner of the sushi bar, normally closest to the doorway, in which to prepare items for the master. He is given a knife to handle. Gradually he is allowed to move closer to the master's position at the bar, the one furthest from the doorway, and is allowed to serve customers and prepare dishes under the strict supervision of the master. At last, after seven years of hard apprenticeship, possibly after sitting a national examination, he can move up to the status of *sushiya* or *itamae*.

The perfect *sushiya* is more than a cook. He is a person who, apart from executing his handiwork perfectly, has total control over a relatively complex operation that consists of preparing sushi while at the same time giving each customer his personal attention, talking with the customer, and being a presence in his environment. A true *sushiya* is a multi-talented, multi-tasking artist and a gifted conversationalist.

Naturally, the training of a *sushiya* is no longer carried out exactly in this way, not even in Japan. Nowadays, and especially in the West, many *sushiya* acquire their skills in schools and sushi academies, and many sushi chefs are self-taught.

PREPARATION OF SUSHI

A Japanese proverb has it that women's hands are too warm to make sushi. Hence, there are only male sushi chefs. This is of course complete nonsense. But it is true that the sushi chef needs an abundance of cold, clean water. Preparation of sushi also requires patience and a certain amount of practice in cutting up fish and shell-fish, in cooking rice, and in designing and shaping the different types of sushi.

HOW IS SUSHI MADE?

Now I am going to run through some of the simplest recipes for sushi, which you can make in your own kitchen. I will talk about how I handle the raw ingredients, the special way sushi rice is cooked, and how to slice the fish. Then I will show you how to make *nigiri*-zushi shaped by hand, rolled *maki*-zushi, pressed *oshi*-zushi, and different types of *temaki*-zushi hand rolls.

You need to know only a few rules. And as I said earlier, sushi is to a great extent a question of a special attitude toward food preparation. When you have learned the rules and practiced them, what counts most is to forget all about rules and principles and let intuition take over. It is just like composing a *haiku*.

I am also going to show some concrete examples of the different types of sushi and how you can prepare, present, and serve them.

An important fundamental principle is to keep your kitchen counter well organized with everything in its proper place. You need to have running cold water within reach and you need to have room in the refrigerator to store fresh raw ingredients and sliced fish until you need them. It is also very helpful to have space to set out the various bowls and plates required for preparation and for serving the finished products. Get your working tools in order and have them all within easy reach.

Those raw ingredients which require a longer preparation time must be made ready before you start. Naturally these include, first and foremost, the ones that need to be marinated, cooked, or frozen and thawed before they can be used. It is also necessary to clean and slice, as appropriate, vegetables, mushrooms, and salad ingredients, and to make *tamago* omelettes. Then whole fresh fish must be gutted, possibly skinned, and cut into fillets of suitable size. If the skin is to be tenderized, this should also be done before the sushi making starts. Shellfish need to be cleaned and removed from their shells.

Typically, three quarters of the time involved in making sushi is spent on getting everything ready.

The next step is to prepare all those items which will not be affected by being made ahead of time. Have the necessary bowls and plates ready for them so that you do not have to handle them more than absolutely necessary.

Vegetables, mushrooms, omelette, and avocado can profitably be prepared first because they are not affected by standing for a while. There can be a problem with sliced avocado as the surfaces can turn brownish, but this can be resolved by sprinkling the cut edges with a tiny bit of lemon juice or, even better, with a little of the marinade from pickled ginger (*gari*). *Wasabi* can also be prepared in advance, bearing in mind that the longer it stands, the less pungent it will be.

Timing is important. This is true for the professional sushi chefs and even more so for amateurs. It pays off to have planned in advance exactly what you are going to make and in which order. When you become more experienced, you can improvise.

Once you acquire a little practice, you can count on spending about two hours to prepare a sushi meal for four people.

学
術 SUSHI BAR SLANG. Sushi chefs often have their own special words for the ingredients used in sushi. Examples of sushi bar slang include *shari* (sushi rice), *gari* (picked ginger), *sushi-dane* or *neta* (that which is placed on top of sushi, for example, fish), and *agari* (tea).

Sushi is prepared in the order in which it is to be eaten, and most preferably one or only a few pieces at a time. If I am making a sushi meal for more than four, I first completely slice fish and shellfish and put the slices (*tane*) in the refrigerator while I shape the rice balls for *nigiri*-zushi. Rice balls made in advance must be covered with a cold, damp cloth so that they do not dry out before use. Do not put the cooked rice or the rice balls in the refrigerator because sushi tastes best at, or slightly above, room temperature.

Even if it seems like a waste of good resources, you have to prepare yourself to use great quantities of very fresh, clean water. The cloths have constantly to be rinsed and the hands cleaned and made wet before you handle the next slice of fish or make a new piece of sushi. Rice, especially, tends to stick to the fingers if they are not damp. Rinse your fingers regularly in water and dip them in a bowl of water that has a little rice vinegar added to it before you pick up rice for *nigiri*-zushi or *maki*-zushi.

Make a habit of rinsing off and drying, on an on-going basis, the tools which you will use over and over, including knives, cutting board, sieves, and bowls. Focus on cleanliness and freshness.

Keep the kitchen counter clean and in order as you work. This is good *wabi*. It is possible that this will take a little longer but, if your experience is like mine, you will discover that the joy in preparing sushi increases in an aesthetically pleasing environment.

Personally, I really derive great pleasure from listening to my favourite music while I am making sushi. It has a calming effect and helps me to get into a good working rhythm.

Finally – do not panic, but savour the moment!

COOKED RICE FOR SUSHI

Properly cooked rice is the determining factor for a successful sushi meal and it is in many ways more important than the fish. For this purpose one must use only short-grained sushi rice. Good quality sushi rice is considerably more expensive than ordinary rice, but it is not worth using cheaper rice. Sushi made with rice that is too hard or has been cooked to mush is a disaster. The rice is cooked to the point where the grains are not completely plumped up and can still absorb moisture when the rice vinegar is added.

As soon as the rice has cooked it must be cooled quickly. Old Japanese recipes stipulate that once the sushi rice is cooked it must be turned out onto a wooden surface and fanned with a special fan (*uchiwa*) to cool it. The best way to cool rice in your kitchen is to spread it out in a wooden tub (*hangiri*) that has first been soaked in water and which will absorb excess moisture, if there is any.

When the rice has stopped steaming, the rice vinegar mixture (*awase-zu*) described in the recipe is added. The finished sushi rice, called sushi-*meshi*, must be allowed to cool some more, possibly covered with a damp cloth. The cooling ensures that the *starch* conversion comes to a halt and that the rice does not overcook. At this time, the rice vinegar also penetrates the grains. Sushi rice grains have to stick to each other, but only just.

調理法 **CLASSICAL SUSHI RICE.** The traditional way of making rice for sushi, sushi-*meshi*, utilizes a special rice vinegar mixture, *awase-zu*, which is added to the cooked sushi rice.

Sushi rice for four

625 millilitres (2 ½ cups) rice

750 millilitres (3 cups) cold water

Rice vinegar mixture (awase-zu)

125 millilitres (½ cup) rice vinegar

22 millilitres (1 ½ tablespoons) sugar

15 millilitres (1 tablespoon) salt

The rice is washed several times in cold water, being ready for cooking when the water runs completely clear. The rice and water are put in the rice cooker or a pot with a heavy lid. While this is happening, soak your wooden tub (*hangiri*), which you will need later, in water to saturate it so that the cooked rice will not adhere to it.

The rice cooker will automatically turn off when the rice is almost done to the right point for making sushi. If you use an ordinary pot (which I find very tricky), you have to allow the rice to cook with the lid on until the water is absorbed, typically 10-15 minutes.

You can make somewhat tastier rice if, during the cooking, you add a couple of strips of *konbu* (kelp) and maybe a teaspoon of *katsuobushi* flakes. You must not wash the strips of *konbu*. Some of the white sediment on the surface of the kelp is *monosodium glutamate*, which imparts *umami* flavour.

When the rice has cooked, it is immediately turned out into the previously soaked wooden tub. The rice vinegar mixture is then folded in carefully with a wooden paddle. This should leave the individual grains with a smooth and glistening surface.

調
理
法
Sushi rice the easy way. Here is a recipe which I have evolved through a process of trial and error and which suits my own taste. Alas, it is a given that the recipe would not live up to the expectations of a professional sushi chef.

Sushi rice for four

625 millilitres (2 ½ cups) rice

750 millilitres (3 cups) cold water

125 millilitres (½ cup) rice vinegar

22 millilitres (1 ½ tablespoons) sugar

15 millilitres (1 tablespoon) salt

30 millilitres (2 tablespoons) rice vinegar powder (*sushinoko*)

The method is the same as for classical sushi rice, with the following exceptions:

Rice and water are cooked together with the vinegar, salt, and sugar.

When the rice has finished cooking, the rice vinegar powder is sprinkled over it and it is turned over carefully with a wooden paddle. This should leave the rice with a smooth, glistening surface.

Cooked sushi rice is cooled in a wooden tub.

The rice must not stick together in a lump, like Arborio rice, and the individual grains of rice should be shiny.

The rice must be cooled before it is used for sushi. I prefer it to be at, or just above, room temperature. It is said that sushi rice should have the same temperature as one's skin. If it is too cold, the taste is less intense. Furthermore, too cold rice has a texture that does not go perfectly with fish in *nigiri*-zushi.

While the rice is cooling after it has cooked, it is important that it is not allowed to dry out and that moisture is present, because otherwise the grains around the edges will become hard. This problem can be solved by covering the wooden tub with a damp cloth. A single hard grain of rice is sufficient to spoil the experience of eating a piece of sushi.

Every serious sushi chef has his own secret recipe for making *awasezu*. The critical item is the sugar content which has to balance the acid vinegar taste. Experienced sushi chefs vary the sugar content in accordance with the season of the year and, in addition, will use a sweeter rice for *chirashi*-zushi and *maki*-zushi than for *nigiri*-zushi.

It can be challenging to cook sushi rice without having a layer of rice grains that have burned onto the bottom of the rice cooker or the pot. These rice grains form a stuck-together, crisp rice cake, which makes a perfect little snack. Provided that the rice grains are not too far gone, the rice cake of 'burned-on rice' is removed, possibly toasted some more, and then cooled. The rice cake can be sprinkled with a bit of lightly salted water and, if desired, with a little cut up, toasted *nori* and served either as a snack before the meal or as a concluding dish.

It is good *Zen* that nothing should be wasted.

CUTTING UP FISH

You will often find yourself in a dilemma when buying fish for sushi. At the fish store you will generally be able to get both whole and filleted fish. It is obvious that it is easier to judge whether or not a whole fish is fresh. If you have confidence in your fishmonger, you can feel quite safe buying fillets or asking him or her to cut the fish up for you. The possibility of making such a choice most often exists in the case of small and ordinary fish, but not in the case of large fish, for example, tuna or other fish that rarely arrive whole at a retail store. You can choose to purchase frozen, sliced fish fillets which have been especially prepared for sushi or have faith in the fishmonger's skill in so doing. It is all a question of trust.

First you have to gut the fish and clean it thoroughly in cold, clean water. For filleting, use a sharp filleting knife. The sushi knife cannot be used for this purpose as it is important to avoid dulling it on fish bones. In principle, fresh whole fish can be filleted in two ways: in three pieces (*sanmai oroshi*), namely, two fillets and the skeleton, or in five pieces (*gomai oroshi*), namely, four fillets and the remaining bones.

珍談 IT IS SAID THAT fish for sushi must be completely fresh, but exactly how fresh? A tall tale is told about the ultimate experience in eating fresh fish as sushi. The chef takes a live fish out of a fish tank which is placed on the sushi bar, slices a piece off for sushi, and returns the fish to the tank. The fish stares resentfully at the customer as he or she is eating the sushi.

Cutting up fish in three pieces (*sanmai oroshi*).

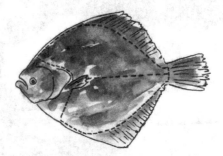

Cutting up fish in five pieces (*gomai oroshi*).

The *sanmai oroshi* technique is used for small round-bottomed fish, for example, salmon, mackerel, and herring. The *gomai oroshi* technique is suitable for flatfish, e.g., turbot, and for large round-bottomed fish, e.g., tuna.

Once the fillets have been cut, the edges are trimmed and all remaining fish bones are removed by cutting them away or by using a tweezer. Feel with your fingertips whether any bones remain.

If the skin is to be eaten, the fish must be thoroughly cleaned, and then scraped and scaled, if necessary, before it is filleted.

For those fish on which the skin is not usable, for example, flatfish, the skin can be separated from the fillet by laying it on the cutting board skin side down. With one hand you take the fish by the tail and with the filleting knife you separate the fillet from the skin. This requires a bit of practice and no skin must remain on the fillet. For fish, such as salmon, which have a layer of fat under the skin, it is removed so that the fatty layer is also cut away.

The skin from some fish, for example, pike-perch and ocean perch, can be eaten but it must first be tenderized. On other fish, such as mackerel, it is the outer, very thin and strong membrane which must be removed. This membrane is inedible because it is tough and, furthermore, it can contain bacteria.

Now the fillets can be sliced with the sushi knife into pieces of *tane* or *gu* to suit the type of sushi you want to prepare.

The fine pieces of fish for *sashimi*, *nigiri*-zushi, and *chirashi*-zushi are made by slicing the fillets crosswise at an angle of 45° in appropriate thicknesses. You execute this manoeuvre in a single stroke by placing the thickest end of the knife blade on the fish and pulling the knife toward your body in a gliding motion. This requires some practice. Use the whole length of the knife in making the cut so that you avoid having to slice back and forth. The outermost edge of the knife blade is used only for very delicate and decorative snips.

It has always been drilled into us that we should slice away from the fingers, but the opposite is the way to do it for sushi and *sashimi*.

珍談 IT IS SAID THAT in days gone by it was a mortal sin for a sushi chef to leave a fishbone in a piece of fish. The negligent chef was expected to commit *hara-kiri*, that is to say, commit ritual suicide by disembowelling himself.

調理法 FISH SKIN IS TENDERIZED by a process akin to blanching (*shimafuri*). This technique can be used for fish that have a skin which is not too thick, such as pike-perch and ocean perch. Place the fillet with the skin upwards in a round-bottomed bamboo sieve (*zaru*), which has been placed with the curved side up in the sink. Place a clean cloth on top of the fillet and pour boiling water over it. Then quickly cool the fillet in cold water so that the flesh of the fish is not affected. The skin has now been softened sufficiently to be sliced through easily together with the underlying fillet to make delicate pieces for *sashimi* and *nigiri*- or *chirashi*-zushi.

Fillet of a red tuna cut and trimmed into a size appropriate for slicing pieces (*tane*) for *nigiri*-zushi.

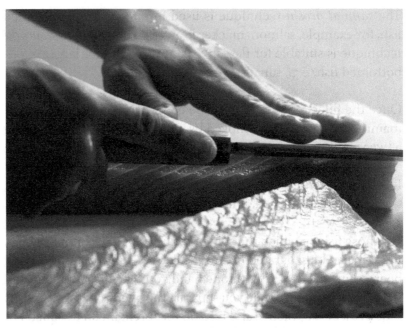

Cutting up of salmon carefully slicing toward the fingers.

Using extreme caution, you must now cut toward the fingers of the hand that is holding the fish. This allows you to support the soft fish and in so doing ensure that the result is a perfect, clean slice, without the fish losing its shape or being crushed in the process. The cut is usually made at an angle, depending on the thickness and the texture of the fillet.

For some fish with a loose muscle structure, especially salmon and tuna, it is important that the cuts in the fillet are crosswise to the connective tissue and not along it. With red fish with a sufficiently firm musculature, such as salmon, you can cut slightly at an angle to get slices with a attractive wavy pattern, which has an fine aesthetic effect.

In the case of certain types of fish, I prefer to leave the skin on the fillet when I slice it. I place the fish skin side down on the cutting board and cut toward the skin, but not through it. This method helps to maintain the shape of the fillet of a fish that has very soft muscles. After the slices have been produced, the individual pieces can be trimmed as necessary to remove the unwanted and slightly tough part of the muscle which is found just under the skin. Turbot, brill, and Baltic whitefish are frequently dealt with in this way.

When the skin is to be eaten with the fish, the fillet is sliced skin side up.

When preparing filling (*gu*) for rolled sushi such as *maki*-zushi and *temaki*-zushi, suitable strips can be cut from the fillet without too much fuss. As the fill in these types of sushi is often partly hidden, you can use more irregular pieces and bits left over from the precision cutting associated with *sashimi*, *nigiri*-zushi, and *chirashi*-zushi.

NIGIRI-ZUSHI – HAND FORMED SUSHI

Together with the *maki* roll, *nigiri*-zushi is the best known form of sushi. It is also the one that allows the greatest scope for experimentation and devising new variations of one's own.

In the main, making it consists of shaping a well-formed rice ball, which sticks firmly together, and then to lay and press a piece of fish (*tane*) on top. The amount of pressure is crucial. It all has to be constructed so that it does not fall apart either when it is served or eaten. But it must not be so compressed that the rice and fish lose softness, texture, and consistency. And the rice grains should be able to fall apart as soon as the piece is safely inside the mouth.

There are a number of classical shapes for the rice balls made for *nigiri*-zushi. Most common is the one called *kushi-gata*, which has a flat bottom, an arched top, and is typically ca. 5 cm (2 inches) long. Some shapes, especially those with sharp edges, are more difficult to make and all require different degrees of pressure from the hands and fingers.

A piece of fish, shellfish, or omelette is placed on top of the rice ball, separated from it by a thin layer of *wasabi*.

When you shape *nigiri*-zushi, it is important to have a bowl with clean water, possibly with a bit of rice vinegar added to it, close by in order to be able continually to moisten the fingers so that the rice does not stick and the fish remains moist. At a sushi bar you will see that the chef, after having dipped his fingers in water, claps his hands forcefully together to get rid of the excess water.

珍談 IT IS SAID THAT the most talented sushi chefs can form rice balls for *nigiri*-zushi so that all the grains of rice lie in the same direction.

Kushi-gata

Tawara-gata

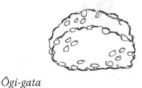

Ōgi-gata

Hako-gata

Funa-gata

The size of the rice ball can vary from sushi chef to sushi chef and the sizes chosen may depend on the manner in which they are to be presented and what they are to be served with in a larger assortment. In some sushi bars I have been served a very large and elongated rice ball with a correspondingly larger piece of fish on it. Such a piece is cut in half in the middle and the two pieces served together with a small decorative space between the halves.

The proportion of the rice ball to be covered by the *tane* is a matter of taste and preference and depends on the thickness and firmness of the piece. Some pieces of fish, for example, salmon (*sake*), are soft and can easily be molded to the shape of the ball. Others, like octopus (*tako*) or omelette (*tamago*), are stiffer and will stick out a bit. You can fasten these firmer pieces of *tane* by folding a narrow strip of *nori* around the whole.

An on-going question is always how much *wasabi* one should put between the *tane* and the rice. Some lean in the direction of not using any because the diner can add *wasabi* to the soy sauce when the sushi is eaten. I do not subscribe to this point of view because the balance in the taste sensation is more perfect when the *wasabi* lies between the rice and the fish and not just on top of the fish or absorbed by the rice ball. I hold that the *wasabi* in the soy sauce should amount only to a fine-tuning.

For *nigiri*-zushi it can be an advantage to prepare the *wasabi* so that it is not too thick. This makes it easier to spread it in a uniform layer. It is important that the *wasabi* not lie in a little compact lump under the fish because this creates a too abrupt and too sharp taste in the mouth. It is not very pleasant to be subjected to a harsh jolt of horseradish smell right up in the nostrils and it diminishes the ability to discern the more delicate taste nuances of the fish.

Nigiri-zushi on which a thin strip of *nori* has been used to secure the *tane*, in this case, a slightly uncooperative shrimp (*ebi*), to the rice ball.

How to make *NIGIRI*-zushi

Take a lump of sushi rice the size of a ping pong ball and press it lightly into an elongated ball in the right hand. Place the piece of *tane* consisting of fish, shellfish, or omelette in the left hand and spread *wasabi* on it with the finger tips of your right hand. Now put the rice ball on top of the *tane* and press both parts lightly together with the left hand. *Nigiri* means to press or squeeze something together with the hands. Then flip the assembled piece over and give it is final shape with the fingers of the right hand. It should be nicely rounded at the ends and stick together properly.

Using a bamboo mat to shape *maki*-zushi. The plastic film serves to prevent the rice from sticking to the mat.

Maki-zushi with complex patterns.

MAKI-ZUSHI – ROLLED SUSHI

I will never forget the first time I saw *maki*-zushi. It was in 1980 at the market on Granville Island in Vancouver, Canada. On the counter of one of the deli stalls there were trays with *tekka-maki*, *maki* rolls with red tuna. I was convinced that it was some kind of sliced-up snake. Since that time *maki*-zushi has become one of my favourite dishes.

Maki-zushi is probably one of the most attractive types of sushi that I know of and there is no end to the possibility of creating variations. In all its simplicity, *maki*-zushi consists of a sheet of *nori* which is rolled together with sushi rice and various fillings, *gu*. The rolls can be thin (*hosomaki*) or thick (*futomaki*), and the *nori* sheet can be on the outside of the rice and the filling or on the inside with the rice on the surface, namely, an inside-out roll (*uramaki*). They can be made with just about anything and you do not need fish, but can instead make them exclusively with vegetarian fillings.

The filling can consist of a single ingredient or several and can be arranged with more or less artistic flourish. It is a form of art to make *maki* rolls which, when they are sliced, display the most beautiful patterns, pictures, or symbols. Here the colour combinations and contrasts play a major role.

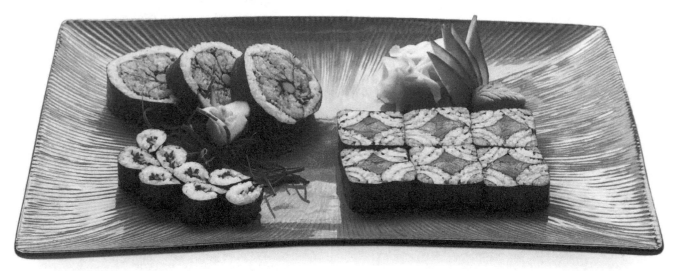

An arrangement of *maki*-zushi consisting of rolls of different shapes and with different internal patterns.

HOW TO MAKE *MAKI*-ZUSHI

Place a sheet of *nori* on the flat side of the bamboo rolling mat (*makisu*) with the shiny side down and one end of the sheet lined up with the end of the rolling mat that is closest to you. A suitable amount of sushi rice is put in a wide strip on the sheet and is then distributed evenly all over the sheet, leaving only about 3 cm (1 inch) uncovered at the end that is away from you.

From this point on it is important to work quickly because the *nori* draws the moisture out of the rice and becomes softer and more fragile as time goes on. First, with your finger tips spread a little *wasabi* on the rice at the end nearest to you. Next, place the filling (*gu*) lengthwise on top of the *wasabi*. Now, holding the mat with both hands, use it to roll the whole works together, exerting a light pressure as you roll forward. At the same time, as the mat is progressively uncovered hold it up and away from the sushi. When you get to the other end, there should be a little overlap where there is only the *nori* sheet. After a few moments, the moisture in the rice will cause the two *nori* surfaces to stick together and the end of the sheet will sit tight against the sushi roll. If you wish, you can help the sticking process along by placing the sushi roll so that the overlap is on the bottom and it will soon form a fine seal. Anything that sticks out at the ends of the roll can be pushed in or cut off. But in some cases the bits that are poking out can have a decorative effect.

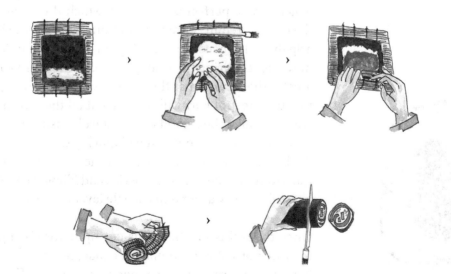

For *maki*-zushi you need to use *nori* sheets, which are available in a number of grades. The finest quality consists of very thin uniform leaves that have a shiny blackish green side. I prefer *yaki-nori,* which is toasted, as it must be completely dry and crisp. Because it absorbs moisture easily, it is stored in airtight packaging. As soon as *nori* is exposed to moisture, it becomes soft and breaks very easily. Therefore, when making *maki* rolls, you have to find a delicate balance between avoiding handling the *nori* sheets unnecessarily with your damp fingers and at the same time keeping your fingers moist while you are placing the rice and filling in the roll.

For the thin *maki* rolls (*hosomaki*) only a half sheet of *nori* is used, while the thick ones (*futomaki*) need a full sheet. I generally use fine, thin *nori* sheets for the *hosomaki*, which are typically ca. 2.5 cm (1 inch) in diameter, whereas I prefer thicker and more robust ones for the *futomaki*, which are about twice as wide. If you feel like experimenting with making very thick rolls, you can use a half sheet lengthwise, if necessary lengthening it with another half sheet. The two sheets can be 'pasted' together by overlapping them slightly and crushing a few grains of rice between them in the seam. Really thick rolls are most easily assembled by rolling them directly with the hands and then using the mat at the end to press them together.

You probably will not be completely satisfied with the end products the first few times you make *maki*-zushi. Making first class rolls, which have a perfect seam and in which the filling lies exactly in the middle, requires practice. But once you get the hang of how to grip the roll, it becomes easy. A good tip is to use quick, determined movements during the rolling process. The more you hesitate, the greater the chance that the roll will end up with ragged edges and that the sheet of *nori* will not make it all the way around. For this reason it might be a good idea to start by using a full sheet of *nori*, even for a thin roll, but you will quickly discover that a roll with a double layer of *nori* is not nearly as delicate. After a while you will learn to adjust the quantities of rice and fillings so that the sheet just goes all the way around in a single layer and has a small overlap.

Once the roll is finished, it is cut up before being served. Thin rolls are normally sliced into six equal pieces, ca. 3-4 cm (1½ inch) long, which are served arranged as a group. The cut is usually made straight across, but a nice decorative effect can be achieved

珍談 It is said that the name of *tek-ka-maki* is derived from the expression for a Japanese gambling joint, *tekka-ba*. The story is that the players did not like to get their fingers greasy while they were playing. Eating *maki*-zushi was an elegant way to solve the problem.

Tekka-maki is *maki*-zushi with red tuna.

by cutting diagonally across every second time. The knife must be completely clean, with no rice on the surface, and it must be moistened with water in order to produce a clean cut.

Thick rolls are cut into thinner slices, about 1.5 cm (¾ inch) thick. Practice is required to make the thick sushi rolls sufficiently firm to hold their shape as they are cut.

The ends of *maki* rolls deserve special consideration. You can either clean them up by trimming them before you begin to divide up the roll or you can leave them as they are and place them end side down on the plate so that they are not seen. It is, however, also possible to use the ends for decorative effect in the best *sabi* tradition. This will make them stand out as slightly odd and imperfect elements in the presentation of the platter of sushi.

You can also give some thought to decorating the ends by placing some filling on the outer edge of the rice before you roll everything together. For example, you could use a little watercress or a small bundle of *enokitake* mushrooms, both of which will look like an attractive tuft when the end pieces are cut off and stood on end. A trick I often use is to place a little lumpfish or salmon roe on the less perfect end piece, sometimes with a little avocado which sticks out a bit and is a good colour complement for the roe.

学術 MAYONNAISE in *maki*-zushi is a Californian innovation which later also became popular in Japan. I cannot stand the taste and the mouthfeel of mayonnaise together with any kind of sushi, so I am going to avoid the subject altogether.

調
理
法

URAMAKI, inside-out rolls, are made more or less in the same way as ordinary *maki*-zushi rolls, and can be both thin and thick. The difference in technique is as follows. The rolling mat is first covered in plastic wrap, which is carefully folded around the edges so that the wrap clings to itself. A few holes are pricked in the film to allow air to escape.

The sheet of *nori* is placed on the mat as if making ordinary *maki*-zushi and the rice is spread uniformly over its entire surface. Then, with a quick motion, you take the sheet and flip it over so that the rice is against the surface of the mat. Because the rice has moistened the *nori* and made it tougher, it is able to keep everything together without breaking during this manoeuvre. The plastic film prevents the rice from sticking to the mat.

Wasabi and filling are added and it is rolled up using the mat. As the rice is now on the outside it is not difficult to make the seam of the roll seal properly. The join can be made almost invisible if the rice has been spread in the first instance right to the edges of the *nori* sheet. Before the roll is sliced it can be sprinkled with, or rolled in, sesame seeds or lumpfish or flying fish roe, to give but a few examples.

It can be difficult to roll a very thick *uramaki* with the bamboo mat. Some sushi chefs boldly make the roll on a damp cloth which gives a better grip on a thick, heavy roll.

When you reach a slightly more advanced stage, you can experiment with pressing the thin *hosomaki* rolls into different shapes. Quite often rolls that are formed in this way are used as filling in really thick rolls, where they can help to create a very special pattern in the slices. For the simplest versions, the roll is shaped so that its cross-section will be triangular or square, or possibly tear-shaped with one sharp edge. When rolls of this type have been sliced, they can be arranged to form a number of decorative patterns, for example, a hexagon or a flower.

A particular type of large *uramaki* roll has a decorative pattern made from individual thin pieces of fish and avocado, or perhaps kiwi, which are placed on top and pressed into the surface of the roll after it has been made. If the topping is very soft, it is helpful to put a piece of plastic film on top of the roll when finally shaping it with the bamboo mat. The roll is then cut leaving the film on to maintain the shape. The film can easily be removed piecewise afterwards.

Some *uramaki* rolls are known as rainbow rolls (*tazuna*-zushi) because the interplay of colours, with stripes of red, green, and white, is reminiscent of a rainbow.

Uramaki – inside out rolls with avocado and grilled scallops, decorated with white and black sesame seeds.

MAKI-ZUSHI GALORE

SAKE-MAKI
Hosomaki sushi with salmon and cucumber.

TAMAGO-MAKI
Hosomaki sushi with omelette.

KAPPA-MAKI
Hosomaki sushi with cucumber. The cucumber can be cut up into either thick square strips or finely julienned slivers (*sengiri*). I prefer the latter because I think that thick cucumber is too hard relative to the soft rice.

TAMAGO-MAKI WITH GREEN ASPARAGUS
Hosomaki sushi with omelette and lightly cooked asparagus.

SHINKO-MAKI
Hosomaki sushi with pickled radish (*takuan-zuke*). The radish is square cut or finely julienned, as for *kappa-maki* above.

SHIITAKE-MAKI
Hosomaki sushi with *shiitake* that have been marinated in soy sauce.

SHISO-UMEBOSHI-MAKI
Hosomaki with a paste made from green *shiso* leaves and minced *umeboshi*.

FUTOMAKI
Giant *maki*-zushi with salmon and green lettuce leaves.

CALIFORNIA ROLL
Giant *maki*-zushi with crabmeat (*kani*) and avocado.

URAMAKI
Inside-out roll with grilled salmon skin and avocado. The roll is decorated with toasted sesame seeds and a little tuft of fine watercress at the end.

GRILLED FISH SKIN

Many of the fish varieties that are used for sushi have a skin which is so thick and tough that it cannot be eaten, so in most cases it is cut off. One well known exception is marinated mackerel, the skin of which is edible after the outer membrane has been removed.

The skin of some fish, such as ocean perch, pike-perch, and sea bass, can be tenderized by blanching (*shimafuri*) once it has been thoroughly scaled. Fish with tenderized skin can be cut up and used for dishes such as *sashimi* or *temaki*-zushi.

In the case of other fish, for example, salmon, the skin can be cut off, scaled, and grilled to be used in preparing *maki*-zushi or *chirashi*-zushi.

Generally speaking, fish skin contains more fat than the fish muscle, often 5-10% more. This is because many of the fat deposits of a fish are found right under the skin in order to function as insulation from the cold water. This layer, called the *dermis*, also contains much connective tissue, that is to say, *collagen*. It is the collagen that makes the toasted skin so crisp and delicious. It is also the substance which, when steamed, gives the fish a sticky, *gelatinous* surface.

Salmon skin is particularly well suited for grilling for sushi. First remove the scales thoroughly. Then skin the fish together with a thin layer of the muscle and connective tissue, so that sufficient fat remains for roasting. The skin is grilled on a warm frying pan, starting with the fatty side. Enough fat will melt out of it to allow you to grill both sides. At the sushi bar you will often see that the chef is holding a piece of salmon skin over an open gas flame to put the finishing touch on it.

The roasted skin is cut up into strips about 1 cm (⅓ inch) in width and placed in a *maki* roll or in a cone-shaped hand roll (*temaki*-zushi).

学
術
OYSTERS (*kaki*). For sushi one uses all the innards of the oyster, namely, the mantle, the gills, the digestive tract, and the adductor muscle. Because of their slimy, limp consistency, oysters are best prepared as battleship sushi.

Oysters from saltier water have a stronger taste. They can keep for up to a week after they are harvested if they are stored in a cold, moist environment, for example, on ice or in a refrigerator either covered with a damp cloth or buried in damp wood shavings or seaweed. The shells should be placed with the curved side on the bottom. As closed oysters contain salt water inside the shell and their cells are filled with substances which must maintain the correct *osmotic* pressure in salty ocean water, one must not store oysters in fresh water or in an ice mixture with fresh water.

GUNKAN-MAKI – BATTLESHIP SUSHI

Battleship sushi is a type of hand formed *maki*-zushi, often made with filling that can be difficult to use on *nigiri*-zushi or in *maki*-zushi because it does not stick together sufficiently or is too soft or too moist. You simply shape a little boat by fastening a strip of *nori* around a rice ball and then load it with the fill. It is also a way to present filling which is particularly attractive or expensive, for example, real caviar or sea urchin roe, *uni*.

Gunkan-maki (also known as *kakomi*-zushi *or funamori*) can have a completely round shape, but the 'boat' is usually a little elongated. The strip of *nori* is fastened tightly around the rice ball, and as a special touch you can let the outer edge of the strip stick out to resemble a little banner.

Battleship sushi with oysters (*kaki*).

Battleship sushi with marinated mackerel (*saba*).

Battleship sushi with salmon roe (*ika*).

Battleship sushi with lumpfish roe.

Battleship sushi with scallop (*hotategai*).

Battleship sushi with flying fish roe (*tobiko*).

HOW TO MAKE *GUNKAN-MAKI*

Cut sheets of *nori* into strips about 3-4 cm (1½ inches) wide and 15 cm (6 inches) long. Shape a little clump of sushi rice into a round or elongated ball and fold the *nori* strip around it. The *nori* strip can be fastened together by crushing a single grain of rice between the end of the strip and the shaped sushi. Place the piece on the cutting board and gently compress the rice so it is below the top edge of the *nori*. The piece should now resemble a little boat ('battleship'), which can be loaded with filling (*gu*) up to the edge of the strip of *nori* and possibly a little higher. On the top of the rice, spread a thin layer of *wasabi* using the tips of your fingers. Then add the filling, which can be finely cut up pieces of fish, roe, oysters, chopped scallops, etc. The finished *gunkan-maki* is often decorated with a little bit of green, for example, cucumber, avocado, or mild watercress.

学術 **Two types of *chirashi-zushi*.** *Chirashi* means to scatter or apportion and there are two ways to make it. One way is by mixing the filling ingredients (*gu*), such as raw or cooked fish, shellfish, and vegetables, with the rice (*gomoku*-zushi). This is an easy way to make a lunch-time meal.

The other way, called Tokyo style, is more elegant and refined. Fish, shellfish, omelette, and vegetables are placed as a decorative cover on top of a lightly pressed together layer of sushi rice.

珍談 **It is said that** Japanese eat the *ebi* first if it is included in a *chirashi*-zushi. This is done out of respect for the shrimp, which can live to an advanced age.

Chirashi-zushi – scattered sushi

Chirashi-zushi is the easiest and fastest type of sushi you can make. In what is known as the Tokyo style it consists mostly of 'spreading' a variety of sushi-*dane* on top of a layer of rice in a flat bottomed bowl. The principal rule is that *chirashi*-zushi consists of two thirds rice with one third filling on the top. *Chirashi*-zushi is a good starting point for someone just learning to make sushi, as many of the ingredients can be prepared ahead of time. It is also a good, stress-free way to prepare sushi when several people are coming to eat a whole meal together at the same time. *Chirashi*-zushi always turns out and it provides you with an opportunity to practice your skills in making decorative garnishes, combining colours, and presenting food aesthetically.

You start by putting a layer of sushi rice in the bottom of the bowl and pressing the rice gently together with the top of the outer part of the fingers of one hand held together. The fingers must be moistened in water so that the rice does not stick to them. The layer of rice should be firm, but not too firm. On top of the rice you can spread a thin layer of roe, for example, *tobiko* or lumpfish roe, or toasted sesame seeds. These will contribute a pleasant, crunchy taste sensation. Using metal or dry wooden chopsticks you sprinkle a thin layer of finely cut up *nori* strips. These should be 3-4 cm (1½ inches) long and a few millimetres (⅛ inch) wide and can be cut from a sheet of *nori* with scissors or a knife. You cannot spread them evenly with your fingers unless your fingers are bone dry and that is why you should use chopsticks. As *nori* absorbs moisture, it is important to work quickly at this point to finish preparing the bowl. Alternatively, you can sprinkle the *nori* strips on top of the *chirashi*-zushi when it is ready to serve.

Now you arrange, as decoratively as you can, the slices of fish and a variety of shellfish, roe, vegetables, and omelette on the layer of rice. It will be more attractive if you utilize several colours, and intersperse some green elements, for example, avocado, cucumber, or green *shiso* leaves. The individual pieces of fish are typically cut into slightly thicker, but smaller, pieces than for *nigiri*-zushi – you can experiment with different sizes and shapes. As a final touch, a small pile of *wasabi* is placed somewhere between the pieces.

Chirashi-zushi is usually arranged in a flat-bottomed black and red lacquered bowl.

Chirashi-zushi with red fish (tuna), white fish (tilapia), shiny fish (mackerel), orange fish (salmon), something yellow (omelette with *nori*), something green (cucumber and spring onion), as well as shrimp, squid, roe, and grilled *tofu*, all placed on a layer of rice in a bowl.

OSHI-ZUSHI – PRESSED SUSHI

Pressed sushi has its origins in the dish that was its prototype, *nare*-zushi, which is cooked rice pressed together with fish. Its descendant, pressed sushi (*oshi*-zushi), has survived into modern times in the Osaka region (Kansai), where the pressed sushi with mackerel (*battera*) is renowned.

It is difficult to make *oshi*-zushi without a good wooden mold (*oshibako*).

The mold is soaked in water so that the rice will not stick to it. It is also a good idea to put a piece of plastic wrap on the bottom piece before the mold is assembled because it makes it easier to remove the finished product. An additional advantage is that you can use it to wrap up the sushi if you need to store it for a while before serving it.

The simplest version is made by putting a layer of rice in the mold with a little *wasabi* spread on top and then placing a mackerel fillet on it. I usually trim the underside of the fillet so that it can lie perfectly flat on the pressed rice. Everything is pressed together by pushing down on the lid of the mold. In principle this is the sum total of a dish of *battera*. You can also make pressed sushi the other way around, starting by placing a piece of fish in the bottom of the mold. This method is preferred if the filling consists of several small pieces, for example, marinated anchovies. But you should note that with the upside down method it can be difficult to control how the desired pattern formed by the fish will turn out.

When the sushi is to be served, the pressed mass is sliced into appropriate size pieces, often ca. 2 cm (¾ inch) wide. The knife must be sharp and it might be necessary to wipe off grains of rice that cling to it as you go along. In order for the slices to have a clean, sharp edge, it is important to use the entire length of the sushi knife when cutting and not to press too hard. Let the knife do the work. If the *oshi*-zushi has rather soft fillings on top, such as avocado, it can be an advantage to leave the plastic wrap in place while it is being sliced. The strips of wrap can be carefully removed afterward.

Oshi-zushi with salmon (*sake*) and avocado.

Oshi-zushi with marinated mackerel (*saba*) and avocado.

Oshi-zushi with shrimp (*ebi*), prepared in a round mold and decorated with a bit of fine watercress.

Many varieties of pressed sushi can be produced by placing successive layers of fish, rice, *nori* sheets, mushrooms, avocado, etc., almost like a layer cake. This is limited only by your imagination. As you become more advanced, you can play around with different colour combinations, which will show up as fine layers when the pressed sushi is sliced. In addition, you can experiment with making squared or interwoven patterns on top, for example, using red salmon, yellow omelette, black *nori*, and green avocado.

Oshi-zushi is an ideal type of sushi to prepare in advance and have on hand, wrapped up in plastic film in the refrigerator. It can be sliced just before being served.

Oshi-zushi can also be served instead of tapas or canapés preceding a non-sushi meal or at a reception.

Cone shaped *temaki*-zushi with sushi rice, strips of cucumber, and red tuna.

学
術   **TEMAKI AT THE SUSHI BAR is** passed over to the customer as soon as it is ready, either directly from the hand of the chef or placed on a plate. When there are several cone shaped *temaki*-zushi, they are often served upright in a little stand with holes that keep the cones together and prevent the *nori* sheets from uncurling.

TEMAKI-ZUSHI – HAND ROLLED SUSHI

Hand rolled sushi is found in a variety of forms, as small packets, as cylinders, and as cones; the latter being the most common.

The cone can be filled with different types of *gu*: fish, shellfish, roe, grilled fish skin, crabmeat, avocado, cucumber, omelette, green salad leaves, etc. You have to be careful, though, that the ingredients are of uniform hardness so that the cone can be eaten without breaking. Sometimes I top off a cone with a spoonful of roe.

A single leaf of green *shiso* folded around the filling in the cone can impart an extra, delicate taste.

Temaki-zushi can also be rolled into a cylindrical shape. To do this, place a small piece of *nori* on the inside going crosswise to the bottom of the large sheet of *nori* to form a sort of U-shaped base. This prevents the filling from dropping out.

Hand rolled sushi is eaten with the fingers. It has to be eaten immediately after it has been prepared before the sheets of *nori* have absorbed moisture from the rice and while they are still crisp and delicious. Hand rolls which have been lying around for too long can easily become chewy because the *nori* has softened.

Temaki-zushi is well suited for making 'do-it-yourself' sushi at the table and, therefore, it is a good way to prepare a meal of sushi for a large group. You prepare all the filling beforehand and put it on the table in bowls and on platters. The *nori* sheets are in a separate container to remain dry and crisp until they are needed. Guests use chopsticks to select the fillings that they prefer and then make the rolls themselves. You will discover that this leads to a very interactive meal.

HOW TO MAKE *TEMAKI*-ZUSHI

Use a half sheet of *nori* per cone. Place it in the left hand and put a clump of sushi rice in the middle of the sheet. Spread a little *wasabi* on the rice. Then put the filling (*gu*) on top of it. The cone is now closed by folding the left side in over the rice and rolling it into a cone. Those with less practice in carrying out this manoeuvre will probably need to call on their right hand for help.

Family sushi or children's sushi
(*kodomo*-zushi).

調
理
法
Children's sushi. The recipe is easy. Cooled sushi rice, a rolling mat, sheets of *nori*, cucumbers, carrots, omelette, avocadoes, and a piece of salmon fillet. Last, but not least, let the children add a good dollop of imagination to set the preparation in motion and it will unfold on its own.

Children's sushi

Children love to make sushi, so this is a golden opportunity to tap into the knowledge acquired in kindergarten about modelling clay and play-dough. Even smaller children can easily put together *maki* rolls and *nigiri*-zushi. In this way they can learn about food that is both healthy and attractive and is even fun to prepare.

In Japan one finds a special children's sushi or family sushi (*kodomo-zushi*), which is sushi with rice but with only a little, or no, fish. But, when it comes to colour and shapes, it is both festive and humorous.

Nigiri rice balls can be pressed into small versions of a host of shapes such as hearts, flowers, or geometrical figures. Toppings are cut to the same shapes and can consist of a layer of yellow omelette, red salmon, white fish mousse dyed with food colouring, and so on. The contents of thick *maki* rolls can be chosen and structured in such a way that when the rolls are sliced colourful patterns, signs, and even faces appear.

My advice for families with children who want to experiment with making sushi is to encourage the children to prepare some of it themselves. Just keep them away from your sharp sushi knife!

Let them have some cooled sushi rice, a rolling mat, some sheets of *nori*, cucumber, avocados, and a piece of salmon fillet. Also, let them experiment a bit with the *wasabi*. You will need to help younger children with the cutting up, but older children can easily handle a not too large or sharp kitchen knife. You will quickly discover that your children's imagination goes far beyond what is found in any recipe which you or I could give them.

My own two children started to make children's sushi when they were under ten years old and derived great pleasure from it. I think this is probably why sushi is still their favourite meal and now, as adults, they have started to make it on their own.

TEMARI-ZUSHI – SUSHI BALLS

A simple, exciting way of making sushi consists of folding and pressing a thin slice of fish, a leaf of green *shiso*, or a slice of avocado around a ball of sushi rice with the help of a piece of plastic wrap. The wrap is twisted around the ball to press the topping onto it and is then removed carefully. The finished balls are decorative and are an excellent addition to the repertoire of sushi for children.

You can make them exceptionally elegant if you place small pieces of something green, for example, avocado or cucumber, under a thin slice of white fish. It will appear as a faint green glow through the somewhat translucent fish.

Temari-zushi in the form of small balls of sushi rice, around which thin slices of salmon have been pressed with the help of plastic wrap.

Temari-zushi made with a white fish (pike-perch).

Bordering on madness

ON WHEN THE HUMAN BEING BECAME HUMAN

The human being has a disproportionately large brain in relation to the body. The brain of an adult accounts for about two percent of the weight of the total body. The same is true for a dolphin. By way of contrast, a large animal like a cow or a rhinoceros has a brain that is equal to less than one thousandth of its body weight. We probably all think that humans and dolphins are smarter than the rhinoceros and that, possibly, this is due to our having a brain that is relatively large in relation to our body. But what about Neanderthal Man? We know that his brain was larger than ours.

The fact of the matter is that it takes more than a large brain to make *Homo sapiens* intelligent and creative. The late English scientist and doctor David Horrobin proposed that the determining factor is the so-called connectivity of the brain. The brain is made up of tightly packed nerve cells; about 100 billion in the cerebral cortex. Thus, the brain has as many nerve cells as the Milky Way has stars. Each of these nerve cells, called neurons, forms electrical connections with a number of other neurons via what are known as synapses. A neuron with its synapses is typically linked to several thousand other neurons. In the cerebellum one finds a cell type that communicates with up to 200,000 other neurons.

It is this connectivity and the concomitant plasticity of the brain that forms the basis for the brain's ability to recognize and store patterns and complex impressions. It may also be the basis for consciousness.

Each synapse consists of a membrane that contains fats, especially the superunsaturated fatty acids arachidonic acid (AA) and docosahexaenoic acid (DHA), which we know from the unsaturated fats in fish muscles. The determining factor in building up the right connectivity of the synapses, which connect the nerve cells in the brain, is precise control over their formation and maintenance. This is true for the growth, breakdown, and reforming of the synapses in their contact points between the neurons. The control is ensured by a set of enzymes, that is to say, proteins, which can build up and break down lipids that contain the special fatty acids, AA and DHA. Lipids are fats that have a head and two tails. The tails are made up of hydrocarbon chains from the fatty acids and the heads are composed of various water soluble chemical groups. Some of the enzymes that act on the synapses, called lipases, are able to cut through the fatty acid chains of the lipid molecules or alter the character of the lipid head. In this way, the lipases can remodel the synapse membranes. Other enzymes, known as acyltransferases, can build up the fatty acid chains of the lipids.

Lipoproteins, a particular type of fat protein, are used to transport the right molecular building blocks to and from the synapses, where the enzymes are active. These transport molecules are of the same type as the ones responsible for the traffic in the body that brings cholesterol from the liver to circulate in the bloodstream. In other words, the lipoproteins deliver the correct fatty acids to the cerebral tissue. The evolutionary determining condition which led to this effective, delicate fat transportation system is in all likelihood the same as the one that brought about the formation in humans of the subcutaneous layer of fat, buttocks, and large female breasts.

These are the particular attributes that distinguish *Homo sapiens* from the other great apes. It is also worth noting that human children are born with significant fat deposits, which ensure that there are sufficient quantities of fats for development after birth.

Horrobin's contention is that if anything goes wrong with this delicate control system, which steers development and maintenance of connectivity in the synapses, it can lead to abnormal brain development and possibly to psychiatric disorders. There are at least two possible causes of such a breakdown and an associated shift in the balance between the fats in the nerve cells. Each of them is attributable to a defect in at least one gene. The first possibility is that too much AA and DHA is released because of overactivity on the part of the lipase enzymes that shear off the fatty acid chains of the lipids in the synapse membranes. The other possibility is that the enzymes that ensure that AA and DHA in the correct proportions are incorporated into the synapse membranes are underactive. Diseases that are attributed to the breakdown of the first type of mechanism encompass the bipolar diseases. If the other mechanism malfunctions, it can lead to personality changes and dyslexia.

The question then arises as to what happens if there is more than one gene defect and both types of breakdown occur in the same person? Horrobin suggests that this occurrence is the biochemical cause of schizophrenia. According to the World Health Organization's standardized criteria for schizophrenia, this disease is prevalent in between 0.5-1.5% of a population group, regardless of gender or race. Some have advanced the claim that schizophrenia is more prevalent in populations with a Western lifestyle and a diet low in polyunsaturated fats. Something in this argument, therefore, points to the idea that diet is as important a factor as genetics in the incidence of some types of mental illness.

Horrobin concluded that the manifestation in humans of the possibility of developing schizophrenia coincided with the evolutionary stage when the biochemical systems that build up, modify, and control connectivity between the nerve cells in the brain were established – that is to say, the point at which *Homo sapiens* became human in same sense as we now understand it. It is supposed that this took place between 50,000 and 200,000 years ago. By then the human brain had been large for a long time, but until this time it had lacked the biochemical systems that would permit the development of a complex brain – a brain sufficiently complex to foster creativity, intelligence, and cultural expression. It was only at this juncture that a breakdown of these biochemical systems could bring about the psychological illnesses that followed in their wake.

In this sense, one can interpret the title of Horrobin's controversial book, *The Madness of Adam and Eve*, as an expression of the correlation between schizophrenia as a disease and the very distinguishing characteristics which make us human.

ARRANGEMENT AND PRESENTATION

Sushi and *sashimi* are arranged on platters and trays, in bowls and on wooden planks, and even in small boats. The combination of the individual items, the harmony of their colours, and the overall presentation are a study in *wabi sabi*. The arrangement of the various elements and their aesthetic appeal are just as important as the food itself. It is not about heaping things on a plate, but rather about respecting the spaces that are left empty.

SUSHI IS LIKE A LANDSCAPE

The arrangement and serving of sushi and *sashimi* are at once easy and difficult. Easy, because you can use simple platters and bowls. It is unimportant if they are worn. Servings often consist of only a few items.

But the difficulty comes in choosing the individual elements and shaping them into a harmonious composite. It is said that presentation of sushi is the last thing that a sushi chef learns to perfection. The approach he takes resembles that of a talented artist when sketching out a painting on a blank canvas.

Seasoned sushi chefs say that sushi must be arranged like a landscape. In the same manner as classical Japanese ornamental horticulture strives to mirror an entire landscape using only a few rocks, an arrangement of sushi must, with only a few pieces, display the qualities of simplicity, ephemerality, and humility. This can be achieved, for instance, by using worn wooden planks or ceramic plates on which the glaze is cracked. It is also done by placing a plain bunch of something green or a simply sliced vegetable on the serving plate together with a few pieces of sushi or *sashimi*. The concepts are very similar to the underlying principles of Japanese flower arrangement, *ikebana*.

In contrast to the focus on symmetry and harmony in the Western world where one uses matching bowls, platters, and plates for a meal, the presentation of a Japanese meal, including sushi, builds on the concepts of asymmetry and contrast. It is immaterial if bowls and dishes are different and do not match, as long as they showcase the food which is presented on them.

EVERYTHING ON A BOARD

At the sushi counter, a simple wooden board with two wooden 'feet' (*geta*) is probably the most commonly used item for serving sushi and *sashimi*. A small wooden boat (*takara-bune*) is another, more sophisticated serving piece for sushi. Black and red lacquered platters and bowls were traditionally used in Japan for serving food. These have been largely replaced by plasticware, although the shapes and the colours remain unchanged. The platters can be simple and rectangular, with either a smooth or a textured surface, but some are fan-shaped to allow for a more interesting display.

Zen inspired rock garden at a Japanese temple.

Sashimi arranged to resemble a miniature landscape on a *raku* platter.

Ceramic and porcelain plates in solid colours or possibly with a classical Japanese motif, like a flower pattern, are also used. A simple bamboo mat is sometimes placed on the bottom of the serving dish.

In Japanese, the overall term for an arrangement of sushi and *sashimi* is *moritsuke* (*moriawase*). In addition, there are many very specialized terms, some of which follow: *hiramori* – thick pieces of *sashimi* standing on edge; *yosemori* – two or three different pieces placed close to each other to form a contrast; *mazemori* – a representative selection of sushi arranged on a small platter; *kasanemori* – slices placed so that they overlap; *sugimori* – pieces placed at an angle to each other in a pile; *chirashimori* – different pieces spread out with spaces in between them.

Selection of sushi in a *mazemori* arrangement.

Sashimi with the fish skin still on is often arranged in a decorative manner by placing the sliced pieces on the fish skeleton (*ikizukuri*) so that they resemble a whole fish.

In order to create contrast (*yosemori*) and special geometrical effects, round pieces of sushi are arranged on plates with straight sides and vice versa. Sushi that is either square or triangular can be placed on a diagonal on a square or rectangular plate. Other means of creating contrast include putting short and long pieces of sushi together and presenting small and large pieces on the same platter.

Sophisticated sushi chefs also make their arrangements in accordance with the seasons. Next time you are at a sushi bar, try to decode your chef's more or less subtle ways of creating a landscape from your sushi or *sashimi*, as well as how the various elements of the arrangement relate to each other.

It is important never to heap too much on the same plate and the bowls absolutely must not be covered with lids. In earlier times, when sushi was something that was sold in street kiosks in Japan, the pieces of sushi were piled on top of each other (*ōmori*). Nowadays sushi is always arranged to resemble a flat landscape. In laying it out, you must regard the empty space as a dynamic place, which frames the food and provides a contrast between emptiness and the objects. What is not there is just as important as what is present. The idea of a simple arrangement is tied to the *Zen* concept of nothingness (*mu*) also known from ornamental horticulture. According to *Zen*, the perfect garden is the garden in which there is nothing more to remove.

Nuances and combinations of colours are also important. In its classical expression, a colour arrangement consists of something white, something red, and something blue or shiny. An example is a combination of white fish, red tuna, and shiny mackerel or herring. Now the colours yellow and orange are also used, represented by ingredients such as omelette and salmon.

Maki-zushi with avocado arranged as a hexagonal rosette.

珍
談
IT IS SAID THAT the reason why *nigiri*-zushi is always served in pairs and never as single pieces or as a threesome is that the Japanese expression for one piece (*hito kire*) sounds like the expression for killing another person and the one for three pieces (*mi kire*) like that for suicide. But Japanese wordplays are subtle and possibly the reason for serving sushi in pairs is solely grounded in a simple Japanese aesthetic. One piece is not enough of a good thing and three are too many; two are just right.

Japanese respect for the mystique associated with numbers is also manifested in the presentation of *sashimi*, always an uneven number of pieces, either one, three, or five. Uneven numbers are positive because they correspond to yang, whereas even numbers correspond to the negative, yin.

Sasa-giri are artistically cut out figures of bamboo leaves (*sasa*) used purely to decorate a serving of sushi and possibly to separate one type from another.

Pieces must always be arranged taking into account visual impact and ease of handling. The most beautiful and most interesting sides are turned toward the person eating, for example, the suction cups of octopus (*tako*) or the fine, crosswise pattern of stripes on the skin of a piece of mackerel (*saba*). Individual pieces of sushi must be placed so that they are easy to take from the platter. *Nigiri*-zushi is turned in such a way that the diner (always expected to be right-handed) can access it directly. A few pieces of *nigiri*, such as shrimp (*ebi*), can be placed crosswise, so that one can better distinguish them from the rest. This is an instance of *wabi* and a breach of the perfect.

Fine patterns and figures (*sasa-giri*) carved from bamboo leaves are used for the decoration and arrangement of sushi. Being practical, the Japanese have now replaced these with a thin green, wavy plastic border, used almost universally in sushi bars. You can also cut out pieces of cucumber or avocado for decoration instead of bamboo leaves or plastic. Finally, green *shiso*-leaves are frequently used to decorate sushi and *sashimi* arrangements.

Wasabi is often shaped with the fingers to form small tops or figures in flower or leaf shapes. This can be done using a mold or a die. In the style called *sugimori*, pickled ginger (*gari*) is stacked up like a small pine cone.

Gari stacked up in *sugimori* style.

Wasabi in the shape of a leaf.

Maki roll with tuna (*tekka-maki*) served as an array of six pieces.

Pressed *oshi*-zushi is sliced for serving and then the pieces are reassembled, possibly with a little distance in between them just so that they can be distinguished from one another. If a whole fish has been pressed into the top, it is vital to preserve the appearance of the whole fish in the presentation.

Hosomaki rolls are arranged in clusters of two, four, six, or eight. You can achieve a special effect by slicing the roll so that a pair of the pieces have a diagonal cross section. The pieces are then placed two by two with the diagonals upward and opposite each other.

You can also serve *hosomaki* by placing some pieces upright and some on their side. In this way it is possible to accentuate the structure and colour combination of the filling in the rolls. *Maki* rolls can be shaped after they have been rolled to be triangular, square, or tear-shaped, rather than round. This allows you to arrange the slices by combining several pieces to bring out different figures and patterns.

Bentō box.

学術 **BENTŌ** is a way of arranging Japanese food, for example, sushi, in a special partitioned box. A *bentō* box most commonly contains rice, *tsukemono*, and other small servings. It is thought that food containers of this type were elaborated by the warlords of earlier times, who used the boxes to ration precisely measured out portions of food for their subjects.

Oshi-zushi with mackerel (*saba*).

"Let food be thy medicine and medicine be thy food"

THE CURATIVE POWER OF FISH

The father of the medical arts, Hippocrates (400 BCE) is quoted as having said, "Let food be thy medicine and medicine be thy food." By this he is expressing the idea that food serves not only as raw material for building up the body and providing energy for its functioning, but should also be viewed as a form of medicine which heals diseases of the body and the psyche. The modern food industry has acknowledged this relationship and is working intensely to develop the so-called 'nutraceuticals', special foods that also have medicinal properties.

We readily accept that food is important for our physical well-being and that a poorly chosen diet leads to lifestyle diseases such as cardiovascular disease, obesity, type 2 diabetes, and cancer. But we have become accustomed to presuming that the nervous system is well protected and that food intake has no appreciable influence on our psyche and on possible psychiatric disorders.

Psychological diseases and mental illness in the West have been increasing markedly and the burden of these diseases is now greater than that of all communicable diseases put together. In Europe alone, the costs associated with neural diseases amounted in 2004 to some €400 billion annually. At this rate, the costs of these diseases are approaching the levels to which cardiovascular disease rose in the course of the 20th Century.

During this period of time, our genes have not been altered, but our eating habits have undergone a radical transformation.

Omega-3 fats from fish and shellfish play a decisive role in the development of the human brain and nervous system. Some experts think that healthy neurological functioning throughout one's life is best maintained by an appropriate balance of omega-3 fats in the diet. For this purpose, fish, shellfish, and seaweed are probably the finest natural nutraceuticals known to us.

Fish and shellfish contain large quantities of the superunsaturated omega-3 fats, especially DHA (docosahexaenoic acid). Omega-3 fats are a family of fats known as essential fats, which is to say that the body cannot construct them from its own chemical pathways and we must, therefore, import them via our food intake. Also, it is only with great difficulty that our bodies can transform other polyunsaturated omega-3 fats, for example, alpha-linolenic acid derived from flax seed among other sources, to superunsaturated omega-3 fats, especially DHA. This is why fish and shellfish are the best source of DHA, while the omega-3 fats from flax seed are a poorer source.

As opposed to the omega-3 fats, the omega-6 fats, which we derive from a variety of vegetable oils and plants such as corn, sunflowers, and soybeans, are abundant in the Western diet, which contains about 10-25 times as much omega-6 as it does omega-3. It is a very unfortunate distribution.

Why is DHA so important? We know that DHA activates over one hundred different genes, lowers the risk of death after a heart attack, regulates the immune system, and improves cognitive abilities in children. DHA is a factor in the development of the nervous system and brain of the fetus and the infant, especially vision and cognitive proficiencies. While it is difficult to study the cognitive development of children, there are indications that it is positively correlated to intake of DHA.

It is noteworthy that the fetus draws heavily on the DHA supply of the expectant mother. In fact, it depletes it to such an extent that children from subsequent pregnancies get less DHA from the mother. It should also be noted that mother's milk is rich in DHA. It is indisputable that DHA is vital in early childhood development. But the DHA content of mother's milk varies greatly, being dependent on the composition of the mother's food intake. In Sudan, mother's milk contains only 0.07% DHA, in the United States 0.12%, in Scandinavia about 0.5%, and in Japan and some coastal areas of China the content can be up to 1-3%.

To put these numbers in perspective, it is interesting to note that the average IQ of the Japanese is about 10% higher than normal. The Japanese typically eat seven times as much fish as people living in the Western world do and their average life span is several years longer.

Dr. Joseph Hibbeln, a psychiatrist and biochemist from the National Institutes of Health in the United States, has studied the way in which omega-3 fats affect the incidence in various countries of bipolar disorders such as schizophrenia and manic depressive illness. His results show that for any given country the prevalence of these diseases decreases as the content of DHA in the mother's milk increases. And the DHA content increases in relation to an increase in consumption of fish and shellfish by the mother. Dr. Hibbeln has also demonstrated that a low intake of fish by pregnant women leads to poorer fine motor and incomplete neurological development in the children. Over an above all of this, it appears that women who eat more than the recommended intake of fish give birth to children who are more advanced cognitively.

Many pediatricians have made up their minds on this issue: they recommend that all women of childbearing age should eat fish, even though the actual scientific underpinnings for so doing have not yet been firmly established and even though it is not yet possible to quantify a recommended daily dosage of DHA. But persuading women to eat fish is very problematic – they have been scared off by publicity about how dangerous it is on account of mercury pollution and toxin accumulation.

American authorities were slow to act on DHA studies. Even though it was known since the 1970s that DHA plays a vital role in child development and that the World Health Organization and the Food and Agricultural Organization as early as 1978 recommended the addition of DHA to infant formula, it was only in 2002 that it was made mandatory in the United States. Experts estimate that American women in all likelihood ingest five times too little DHA and that the most common source of DHA for women is not fish but eggs, which have a reasonable DHA content. Special omega-3 enriched eggs, produced by allowing the hens to eat flax seed, can have an appreciably greater DHA content.

With the mention of DHA enriched eggs, the discussion comes full circle and back to Hippocrates, who also famously said "Let them eat flax!" But probably Hippocrates was thinking more about avoiding constipation than about guarding against cardiovascular disease, cancer, and psychological illnesses.

SUSHI À LA CARTE

*With a just-yanked
radish
pointing the way*
Kobayashi Issa. (1763-1827)

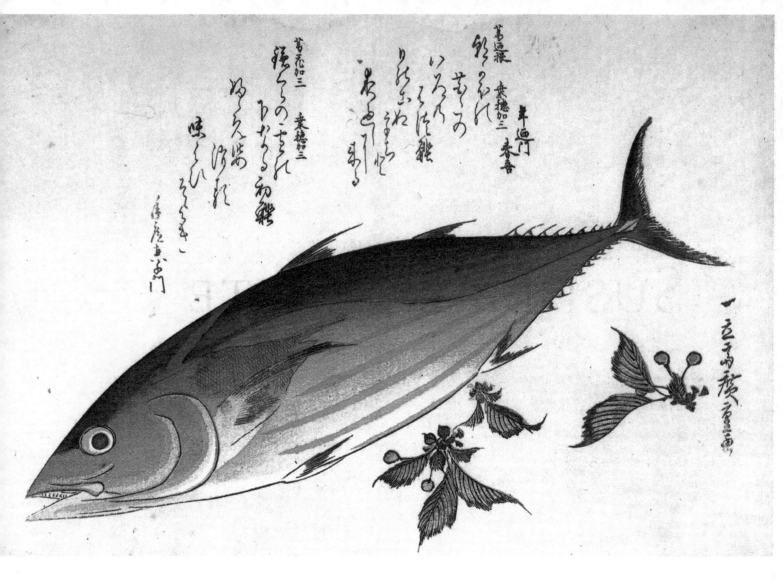

MAINSTREAM SUSHI

There are many different types of fish and shellfish which one can easily obtain, either fresh or frozen, among them salmon, flatfish, tuna, shrimp, and roe. Together with vegetables, mushrooms, and eggs, this list of ingredients makes up the basis for the most common *tane* and *gu* for sushi. They are widely available almost all year round. Many of these ingredients can be used to make more than one type of sushi and *sashimi*.

SALMON (*SAKE*)

Fresh salmon is always good for sushi, especially the fatty part of the fillet – in general, the fattier the fish, the better the taste. There can be a great variation in the flavour of salmon from different waters, with wild salmon often being tastier than that raised in fish farms. Pieces of salmon for *nigiri*-zushi can most easily be cut from a whole, trimmed fillet, but salmon steaks can also be used. Be sure to ask for those that have been cut farthest from the tail.

Avoid slicing the salmon lengthwise along the muscle fibres (*myotomes*). Instead cut the fillet at an angle so that it results in a crosscut with a fine, wavy pattern formed by the white connective tissue. Trim away the dark muscles which lie near the skin, especially along the side. These trimmings can easily be used in a soup. Salmon for *nigiri*-zushi is normally cut in thicker slices than other types of *tane* because it has a very soft consistency.

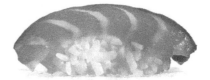

Nigiri-zushi with salmon (*sake*).

Some sushi bars use smoked salmon for *nigiri*-zushi. In this case the slices must be thinner than the ones for fresh salmon. Although I am personally very fond of smoked products, I do not think they go with sushi. The smoky taste detracts from the flavour nuances of the rice. If one absolutely must incorporate smoked fish into sushi, it should be eaten last.

学術 **SALMON DOES NOT** make an appearance on the classical Japanese list of the red ingredients (*akami-dane*) used for *edomae*-zushi, even though it belongs to this colour group. It is said that the reason is that salmon in Japan can be caught only off the shores of the northern island Hokkaido, which in earlier times was regarded as a rather backward place inhabited by dullards. In recent years salmon has become a popular sushi fish in Japan, but most of it is imported from Atlantic aquaculture sources.

Salmon (*sake*) *sashimi*.

Tuna (*maguro*)

Many people regard tuna as the king of sushi. The tuna family consists of at least fifty different species, of which about ten are used for sushi. The bluefin tuna (*hon-maguro*) and yellowfin tuna (*kihada*) are especially well-suited for making sushi. Tuna can live for up to 30 years and grow to great size, up to 3-4 metres in length. The price commanded by a fine tuna can exceed that of the highest quality beef. Tuna can be caught year-round in different parts of the world.

Bluefin tuna, which is about five times as fatty as the yellowfin and consequently has a stronger taste, is regarded as the best for sushi. Tuna has about ten times as much fat in the belly muscle as in the back muscle. The latter are, therefore, red, whereas the former are paler and have less firm layers of connective tissue.

Fish stores do not always have fresh tuna on hand, but frozen fillets can certainly be used, even if they might have lost a little moisture and taste. For *nigiri*-zushi you should make sure that you purchase a fillet that is sufficiently wide to permit you to cut off pieces that are an appropriate size for placing on the rice balls.

学術 **Tuna are predators** which can weigh up to almost 700 kilograms and swim with a speed of 70 kilometres per hour. They can breathe only when water is forced through the gills as they swim. For this reason, they are in constant motion (just like mackerel), ranging over vast stretches of the ocean in their hunt for prey. As tuna are at the top end of the food chain, they can be loaded with environmental toxins. The two principal species used for sushi are bluefin and yellowfin tuna.

Frozen tuna at the fish market, Tsukiji, in Tokyo.

Make sure that you trim away any connective tissue membranes that might be found between the larger muscle fibres. Cut the muscle fibres crosswise or at an angle so that the slice does not come apart along the *myotomes*.

The red muscle bundles of the tuna are surrounded in several places by looser layers of muscle that are richer in fat and serve as insulation. Sushi lovers regard the fatty belly muscle (*toro*) of the tuna as a particular delicacy. It is very expensive, but it melts in the mouth and is often the first piece eaten at a sushi meal. As *toro* is soft and has loose fibres, it can easily fall apart when it is sliced, so this must be done extremely delicately. When tuna is eaten as sushi or *sashimi*, the amount of soy sauce used should be decreased as the fattiness of the tuna increases.

You can present the red tuna in an especially elegant way by preparing it as *tataki* – sear the fillet very quickly on all sides on a frying pan and then slice it. The red colour of the tuna stands out against the brown border that is formed because heating *denatures* the *myoglobin proteins*, which otherwise give the fresh muscle its red colour.

学術 **DOLPHINS ARE UNFORTUNATELY** all too frequently caught up in tuna nets. So to protect them ask your fishmonger whether the tuna is from an area where they are caught using lines and hooks. Regrettably, this method of catching them is not without its problems either, as giant albatrosses often bite into the hooks when they dive for food.

Nigiri-zushi with red tuna (*maguro*).

Nigiri-zushi with fatty tuna (*toro*).

珍談 It is said that the most expensive fish in the world was a 202 kilogram bluefin tuna which was caught in January 2001 and auctioned at the Tokyo fish market for $173,600 U.S., or about $860 per kilogram. It was probably reduced to many hundreds of servings for well-heeled sushi enthusiasts.

Nigiri-zushi with red tuna (*maguro*) prepared as *tataki*.

Mackerel (*saba*)

A fat mackerel is one of my favourite fish for *nigiri*-zushi. The mackerel is from the same family as the tuna and is also a predator always in motion. Its muscles are of the slow variety; they are dark and have a strong, slightly metallic taste. The mackerel, which makes its home in the Mediterranean and the North Atlantic, can attain a size of up to 40 centimetres in length and weigh as much as a kilogram. The substantial *enzyme* content of mackerel causes the dead fish to start to decompose very quickly if it is not immediately cooled on ice.

Fat mackerel caught in autumn are best for sushi. As mackerel can have natural parasites, they cannot be eaten raw. They must be marinated, skinned, and preferably also frozen before being used for sushi. A simple curing technique (*sujime*), involving salt and vinegar, is used to prepare the mackerel, thereby killing the parasites and imparting a firmer texture to the otherwise soft fillet. Marinating also mellows the taste of this very oily fish, which late in the summer can have a fat content of about 20%.

Fillets of marinated mackerel (*saba*).

MARINATING MACKEREL

Clean and gut the fresh mackerel. Then cut the two side fillets free from the bones, using the three-pieces technique (*sanmai oroshi*) that produces two fillets and a skeleton. Any residual bones in the fillets are cut out or removed with tweezers. As mackerel muscle is very soft, you should not attempt to remove the smaller bones which are well anchored in the skin. These bones will dissolve or become very tender during the marinating process and will not be a bother later on.

The fillets are then salted with a thick layer of fine or coarse salt and placed in a flat-bottomed elongated dish. Be sure that the entire surface of the fillet is covered with salt. Let the mackerel stand in a cool place for 2-3 hours, depending on size. The warmer the weather, the shorter the time required for this step. Remove the fillets and rinse off the salt together with the oil that oozed out of the fish. They are then returned to the cleaned dish and vinegar is poured over them so that they are completely covered. Add a tablespoon of sugar and let the mackerel marinate for 20-30 minutes, depending on size.

Remove the fillets and dry them with a clean cloth or paper towelling. Place them on a cutting board and trim the edges with a sushi knife so that the base of each fin is completely removed.

At this point, pull off the outer, very thin and transparent, but very strong membrane. Start by grasping the top part at the thicker end. In most cases, the membrane can be peeled off in a single piece without breaking, but otherwise it will take several attempts to remove it all. Unavoidably, some of the inner, pigmented skin on the fat belly side of the fillet will come off with the outer skin. Because marinating has caused the *proteins* in the muscle to *denature*, resulting in a firmer fillet, you can try gently to remove any remaining bones.

Before use, it is preferable that the marinated mackerel should spend 24 hours in the freezer. Because mackerel is an oily fish, it does not keep for a long time in the freezer. It loses its delicate taste and the fish oil seeps out after about a month's time.

Nigiri-zushi with mackerel (*saba*).

學術 MACKEREL SKIN IS SMOOTH and has a silver sheen. All bony fish have two layers of skin, an outer layer (*epidermis*) and an inner layer (*dermis*). The dermis is related to the connective tissue between the muscle fibres and contains cells with a pigment that is actually the crystallized *nucleic acid guanine*. These crystals impart a silvery white, shiny tinge which is also found in herring. This metallic layer is very prominent when the outer layer of the skin has been peeled off. The glistening effect of the skin acts as a camouflage to help the fish evade predators when swimming near the surface of the water.

Nigiri-zushi with tilapia.

The marinated mackerel fillet is sliced into suitably sized pieces by cutting it diagonally with the skin side up. As the muscle of a fat mackerel can still be quite soft even after it has been marinated, you must slice with a light touch. It can be an advantage if the fillet has not defrosted completely when you slice it. Be sure to cut the slices in such a way that the magnificent pattern of blue-black and dark green stripes in the glistening silver skin stand out. As leaner mackerel can be a little stiff and firm of flesh, it might be difficult to get the slices to adhere to the rice ball. In such a case, you can secure them with a little strip of *nori*.

In many sushi bars a little lump of finely ground fresh ginger or a bit of minced spring onion is placed on *nigiri*-zushi with mackerel. The idea is to create a contrast with the oil-rich fish. On occasion, a sushi chef will allow mackerel to be accompanied by a slice of lemon, presumably because some take pleasure in attenuating the slightly rich taste of the mackerel with the tartness of the lemon. Personally, I would never dream of letting lemon spoil the superb taste of mackerel.

TILAPIA

Tilapia is a fresh water fish which originates in Israel and Africa, where it is found, among other places, in Lake Victoria and the Nile. It is a fast growing fish and it is now raised commercially in many places in the world where the water is sufficiently warm. Tilapia does not belong in the Japanese kitchen and has no common name in Japanese.

Tilapia has fine white flesh with a slight rosy tinge. As tilapia is now one of the top three farmed fish (with salmon and trout), it is widely available where fish is sold, either frozen or fresh. It can be used to make sushi and *sashimi*. It should be noted, however, that farmed tilapia has turned out to have much less omega-3 fatty acid than wild tilapia.

FLATFISH (*HIRAME* AND *KAREI*)

White flatfish are suitable for both sushi and *sashimi*. I know of nothing better than being so lucky as to get a beautiful, fresh turbot in the fish store and then going straight home to cut it up for *sashimi*. Both turbot and brill, which are half-oily fish, make a good, very tasty topping for *nigiri*-zushi. They have a fine texture, especially turbot, and are very succulent. Turbot and brill are best in the northern winter months, especially January and February.

Greenland halibut, which is an oily flatfish, can also be used, but it has a much less delicate taste. Other, smaller types of flatfish such as lemon sole present sushi possibilities, but are rather difficult to slice into suitable pieces. Halibut (*ohyō*) is a very large, lean fish which is particularly good for *sashimi*.

Flatfish are cut up using the five-pieces technique (*gomai oroshi*), that is, four fillets and a skeleton. Pieces for *nigiri*-zushi are sliced from the fillet diagonally so that they are of the right size. As these flatfish have a firm texture, they are easy to cut up, except toward the very ends. The muscle can be somewhat tough near the skin, so you should avoid cutting too close to it. Trimmings and chewy bits can easily be used for soup stock.

Cutting up flatfish for sushi can, however, also be a bit tricky, as the pieces must be neither too thick nor too thin. If they are too thick, the result is a chewy and poor mouthfeel. If they are too thin, the texture of the fish is lost.

For *nigiri*-zushi made with the finest pieces of flatfish, it is considered a particular refinement if the green *wasabi* can just be made out under the pale fillet.

Flatfish contain few *enzymes* in the fillet and, hence, they keep well in refrigerator and freezer.

学術 **FACING LEFT OR FACING RIGHT?** At a certain point in their development, flatfish turn their heads either to the left or to the right side. The fish is no longer symmetrical. The underside becomes white and the topside takes on a darker, camouflaging colour. For a given species, most individuals have turned their heads to the same side.

The Japanese make a distinction between the flatfish that have their eyes on the left side (*hirame*) and those which have them on the right side (*karei*). *Hirame* includes fish such as turbot and brill, and *karei* can be lemon sole, Greenland halibut, or halibut.

Nigiri-zushi with turbot (*hirame*).

Nigiri-zushi with herring (*nishin*).

HERRING (*NISHIN*)

Japanese varieties of herring are used for *nigiri*-zushi, but not very frequently, as they do not keep well. On the other hand, herring roe (*kazunoko*) is a sought-after delicacy for *gunkan-maki*. Herring should not be eaten raw and, therefore, they have to be marinated before they are made into sushi. The herring must be scaled meticulously.

I find it hard to marinate herring so that it turns out well for making sushi. What I do instead is buy herring fillets that have been lightly salted and soak them well before I use them for *nigiri*-zushi. Preserved or soured herring, e.g., 'matjes', are often too salted or marinated with onion or spices which do not go well with sushi. Moreover, preserved herring fillets are usually skinned, resulting in the loss of their beautiful shiny silver skin.

Nishin should not be confused with *kohada* which is gizzard shad, a relative of the sardine.

Nigiri-zushi with Japanese sea bass (*suzuki*).

JAPANESE SEA BASS (*SUZUKI*)

The Japanese consider sea bass to be one of the best fish for sushi. The sea bass wanders from fresh water habitats to the open sea; when it reaches a certain size the Japanese call it *suzuki*. Its flesh is fine textured and white with a rosy glow. Sliced into delicate morsels is it also delicious as *sashimi*.

Sea bass is seasonal, with the quality falling off toward winter as the fish loses its fat. Keep on the look-out for it and you may be fortunate enough to find it fresh, albeit at a high price. Frozen Japanese sea bass may be found from time to time internationally. No exact equivalent to the Japanese sea bass is widely available elsewhere, but a number of similar fish, such as farmed striped bass, are sold as sea bass and found in sushi bars in North America.

PIKE-PERCH

Pike-perch is a fresh water fish which can be turned into very good sushi and *sashimi*. It resembles a cross between a perch and a pike, hence its name. As it is not common in Japanese cuisine, there is no word for it in Japanese. It is a very lean fish which keeps well when frozen.

Pike-perch fillets are cut using the three-pieces technique (*sanmai oroshi*), that is, two pieces and the skeleton. Slices for *nigiri*-zushi are cut from the fillet going in toward the skin. The flesh is completely white.

The skin of the pike-perch, which is beautiful with greeny-brown cross stripes on a white background, is edible if it is first tenderized.

Nigiri-zushi with pike-perch.

OCEAN PERCH

Ocean perch is a deep water ocean fish as is evident from its large eyes. It makes good sushi and *sashimi* on a year round basis and, because it is relatively lean, it freezes well. Ocean perch is not the same as red snapper, which again is not quite equivalent to the sought-after Japanese *tai*, which is a red seabream. Like pike-perch, ocean perch is not traditionally used in Japan and hence, there is no Japanese name for it.

Ocean perch fillets are also cut using the three-pieces technique (*sanmai oroshi*), resulting in two fillets and a skeleton. Slices for *nigiri*-zushi are cut from the fillet in toward the skin. Because of the slightly tough texture of the fish, *tane* is sliced thinly. Its flesh is a pale rose colour, which looks beautiful in an arrangement of sushi.

The skin of the ocean perch is thick and tough, but it can be eaten if it is tenderized.

Nigiri-zushi with ocean perch.

Nigiri-zushi with yellowtail (*hamachi*).

YELLOWTAIL (*HAMACHI*)

Yellowtail, also called Japanese amberjack, is named, as might seem obvious, after the colour of its tail. In Japan, the term *hamachi* covers a number of different species and it is also known by a host of regional names. To give just the most prominent examples, in Tokyo, it is called *inada*, and in Osaka, *hamachi*. To complicate matters further, these overlap with the names used to indicate the size of the fish, that is, *inada*, *hamachi*, and *buri*, as the fish progresses from adolescent to very mature. A close relative of *hamachi* is called *kanpachi*, which is less oily and has slightly darker meat. Regardless of what it is called, yellowtail is one of my favourites for making both sushi and *sashimi*.

Hamachi has an unusual buttery, but firm, texture. It is the fattiest of the white fish used for sushi. In contrast to tuna and other fish, the dark lateral line of the *hamachi* is not trimmed away. On a very fresh fish, this line is an intense red colour and this gives a beautiful effect on *nigiri*-zushi.

EEL (*UNAGI* AND *ANAGO*)

Nigiri-zushi made with grilled eel is a delicacy in Japan. Both fresh water eel (*unagi*) and the slightly larger and leaner salt water eel (*anago*) are used. *Anago* is meatier than *unagi* and contains up to a hundred times more *vitamin* A than other fish.

The eels are not eaten raw, but are steamed after filleting and placed for half an hour in a reduced eel broth to which soy sauce, *mirin*, and sugar have been added. For *nigiri*-zushi, the eel fillet is sliced into appropriate pieces that are warmed, grilled, and placed on the rice ball. The eel is then brushed with a special sweet sauce (*nitsume*), made from reduced eel broth, sugar, and soy sauce.

Nigiri-zushi freshwater eel (*unagi*).

POLLOCK

Pollock is the common name for two species of a lean fish from the same family as the cod. One species has lovely white muscle meat, with a fine, mild taste, which is ideal for both sushi and *sashimi*. But the other, sometimes referred to as saithe, has darker meat with a taste that is too strong for our purposes. Pollock must be frozen before it can be made into sushi.

Pollock can be used for sushi from the summer through mid-winter, but it is best when the weather is colder.

Nigiri-zushi with pollock.

BALTIC WHITEFISH

If you ever have the chance to eat sushi in either Finland or Sweden, you must try Baltic whitefish, which the Finns call 'siika'. Baltic whitefish is a fresh water fish, but it is also found in coastal waters where the water is cold and rich in oxygen. As the name indicates, whitefish is a smooth, shiny fish with fine white flesh.

The siika fillet is very soft and rather difficult to slice. It is terrifically well suited to curing in the traditional Swedish manner, like 'gravlaks'. The salt used in the process helps make the muscle firmer and more congealed, leaving it much easier to cut up for sushi and *sashimi*.

Nigiri-zushi with Baltic whitefish ('siika').

Baltic whitefish ('siika') from Finland.

Nigiri-zushi with giant shrimp (*ebi*).

SHRIMP (*EBI*)

The term shrimp covers a variety of species, found in waters worldwide. The cold water ones tend to be small and are not suitable for sushi, while the larger species found in tropical waters are. Of these, use only the white and pink species for sushi, leaving aside the brown shrimp, which tastes too strongly of iodine. You might look for the ones labelled giant shrimp or tiger prawns. While most shrimp sold nowadays are farmed, wild ones have much more flavour. The best I have ever had for sushi were wild Venezuelan shrimp.

These large shrimp cannot be eaten raw, but must first be cooked. On cooking, the otherwise translucent tail muscles turn white and the blue-green shells become red. The red colour is due to the denaturing of the protein complex, *crustacyanin*, which releases the pigment *astaxanthin*. Shrimp are often frozen after they have been cooked. It greatly improves the taste of the cooked shrimp if you let them stand for a while in a little rice vinegar to which a bit of sugar and *mirin* have been added. This removes the somewhat flat taste often associated with cooked shrimp.

学術 **WITH OR WITHOUT THE HEAD?** Shrimp and other crustaceans have one organ in what we call the head. The organ is a type of liver which secretes digestive *enzymes*. Right after the shrimp has died, these enzymes begin to decompose it, especially the muscle meat in the very desirable tail. For this reason, shrimp are usually sold either live with the head on, fresh without the head, or cooked or frozen with the head on.

When it is cooked, the shrimp's muscle curls up. But a tightly arched shrimp tail is not attractive and it is difficult to place it on a rice ball for *nigiri*-sushi. To prevent curling, you can place a little wooden skewer under the shell just at the bottom of the tail before it is cooked. After cooking, remove the skewer and carefully peel away the shell. Using the sushi knife, make a fine incision all along the bottom of the tail. On the backside of the shrimp there is a black vein which is the end of the alimentary canal. It can contain bits of undigested material and particles of sand. The vein is removed either by pulling it out or by scraping it away with the tip of a knife. You can then press the whole tail flat on a cutting board, making it easy to position on top of a rice ball. Be sure that the rice is pressed firmly against the shrimp, allowing you to fold it over the ball and letting it resume some of its original curved shape.

Ebi are eaten as sushi especially for their texture, which is firm and slightly crunchy. This forms a nice contrast with the softer pieces of fish, not least in *chirashi*-zushi.

学術 **SHRIMP AND THE TASTE OF IODINE.** Shrimp feed on algae and seaweed, both of which contain considerable quantities of iodine and bromophenols. These substances give shrimp, especially wild shrimp, their characteristic taste of the fresh ocean.

調理法 **COOKING GIANT SHRIMP** for sushi. Place the unshelled, skewered tails in a pot with boiling water to which, if desired, a little salt has been added. They should boil for under a minute and then immediately be cooled in cold or ice water before they are peeled. The cooling helps to fix their red colour.

As a special touch in presenting sushi with large shrimp, you can leave the shells on at the very ends of the tails and spread them out like a little fan. You must, however, remove the small sharp part right in the middle of the tail end. True sushi connoisseurs eat these shells with the shrimp.

Nigiri-zushi with octopus (*tako*).

調
理
法
Cooking octopus. Loosen the arms from the body of the octopus, wash everything thoroughly to remove any bits of dirt, and rub the octopus with salt. Place it in lightly salted boiling water that has a bit of vinegar in it. Let it cook for about 10 minutes, until you can easily insert a sharp knife in the arm. This is a real balancing act: if the octopus is undercooked, it will be tough, and if it is overcooked, it will also be tough! The cooking time has to be just short enough so that the *proteins* in the muscle are not *denatured* completely, but the *collagen* in the connective tissue is made sufficiently soft.

Octopus (*tako*)

Octopuses (*tako*) can often be bought fresh at the fish store, but they are also available frozen. They have two rows of suckers on each of their eight arms, which are connected in toward the body by a web. The suckers can contain unwanted material and must be cleaned thoroughly.

As octopuses may have natural parasites, they cannot be eaten raw and must be cooked first. Only the arms are used for sushi and *sashimi*. Cooking makes the soft arms firmer and their surface changes colour from greyish white to reddish violet. This change of colour is attributable to the same chemical process that turns cooked shellfish red. The muscle meat itself becomes completely white when cooked and acquires a slightly nutty taste.

The web is cut away from the cooked arms, which can then be sliced with the sushi knife. The thickest arms are most suitable for sushi. *Tane* for *nigiri*-zushi is cut diagonally across the arm and it is important that the cuts are placed so that the suckers are prominent on the edge of the slices. Many sushi chefs carry out this operation with a slightly choppy motion so that a decorative wavy pattern is formed on the side. The pattern also helps the piece of *tako* to sit more securely on the rice ball. For *chirashi*-zushi, on the other hand, *tako* is cut straight across.

Cooked octopus arms can be frozen for later use. As the muscles have a very low fat content, they keep in the freezer for several months. Some people think that a pre-frozen octopus is more tender after cooking than a fresh one.

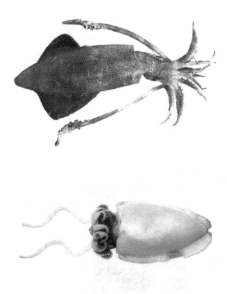

Two types of ten armed cephalopods (*ika*);
Loligo on top, *Sepia* at the bottom.

Nigiri-zushi with squid (*ika*).

CUTTLEFISH (*KŌIKA*) AND SQUID (*IKA*)

In contrast to the somewhat lumpy octopus, the cephalopods with ten appendages are more streamlined and have eight short arms in addition to two very long tentacles used for catching prey. The two main groups of ten-armed cephalopods are the *Sepia* (cuttlefish, *kōika*), which have a plump body and a calcium cuttlebone under the entire back, and the *Loligo* (squid), which have a narrow chitin quill running the length of the back. Both types are used for sushi and *sashimi* under the designation *ika* in Japanese.

The muscle meat of the different species of *ika* is of varying degrees of firmness. Some are thin and soft and others thicker and tougher.

Ika is usually readily available in fish stores, either fresh or frozen. *Loligo* can be eaten raw, while *Sepia* must first be lightly cooked. It is the body, and not the arms, that is used for sushi. The arms can instead be grilled or marinated to make a salad. After shrimp, *ika* is the sweetest type of shellfish used for sushi.

To prepare cuttlefish and squid for sushi remove the arms and the innards from the body. Next, pull out the stiff cuttlebone or quill. Then slice the body open on one side and pull off the grey skin and, with it, the two lateral fins. Only the cleaned tube of the body remains and it is now sliced into pieces of appropriate size. Place the pieces on a cutting board with the outside up. As it is tougher than the inside, use a sushi knife to score it with a delicate diamond pattern to make it easier to chew. If the cuttlefish or squid is quite small, this is a bit of a tricky operation, as it is important to avoid cutting all the way through. You can also make a simpler pattern of parallel lines. In either case, the pattern will stand out as a decorative touch in your *nigiri*-zushi presentation. In addition, the scoring makes it easier for the soy sauce to adhere to the finished piece.

Because slices of *ika* are smooth, they can be difficult to fasten to the rice balls. It helps to hold them in place with a strip of *nori*.

SCALLOP (*HOTATEGAI*)

In contrast to most other species of bivalves, scallops have only one adductor muscle, but it is large and makes up about three-quarters of the total weight. The strong adductor muscle is necessary because scallops are also one of the few species of bivalves which swim about freely. In order to swim, the scallop quickly slams the two shells together and then opens them again very slowly. The adductor muscle consists of fast muscle, which is very tender and it is the only part of the scallop used for sushi and *sashimi*. It has a very sweet taste with barely a hint of iodine. The sweetness is attributable to the large quantity of the *amino acid glycine* and of *glucose*, both by-products of the breakdown of the *glycogen* in the fast muscle. In principle, the scallop roe (the coral) could also be made into sushi but I have never seen it served in a sushi bar.

Shell and innards of a scallop (*hotategai*). The adductor muscle is the round, white shape.

Scallops are both harvested from the oceans and, increasingly, raised commercially in aquaculture settings. In some parts of the world, fresh scallops are readily available, while in others only frozen ones are easy to come by. Frozen scallops should be defrosted slowly in the refrigerator but will, unfortunately, lose some of their liquid. They are well suited for both *sashimi* and sushi. Small ones can be used in *gunkan-maki*.

For *nigiri*-zushi, slice the adductor muscle of the scallop across the grain of the muscle fibres, about three-quarters of the way through. Then fold it open so that the uncut part stands out as a raised section on top of a piece of *nigiri*-zushi. Very large scallops can be cut crosswise to make flat slices which can be fastened to the rice ball with a strip of *nori*.

When sushi is made with scallops it is best to omit the *wasabi*.

学術 SCALLOPS ARE SOLD without shell. Most bivalves, for example, oysters and mussels, can close their shells completely and can thereby preserve their water content even when out of water. Scallops are not able to do this. When they are removed from their element, their water content and dissolved substances will run out and the scallops will die and decompose rapidly. For this reason, the adductor muscle is cut out of the scallop shell quickly after it has been caught and consumed fresh or frozen for storage.

Nigiri-zushi with scallops (*hotategai*).

Nigiri-zushi with omelette (*tamago*).

調
理
法
OMELETTE (*tamago-yaki*). Crack three eggs open in a small bowl. Add a little each of salt, sugar, and *mirin* (optional), and beat all together lightly with a fork. Heat a pan greasing it with the tiniest amount of fat – its taste must be virtually undetectable, so apply it to the pan with a piece of paper towel. Pour the egg mixture into the pan at low heat a little at a time until it gradually sets. The egg mixture sets because the *proteins* in the egg yolk and white *denature*. With a set of chopsticks or a wooden spatula fold the set egg mixture together on itself several times until it gradually takes on the shape of a box.

It is an advantage to use a small rectangular pan (*tamago-yaki-nabe*) because the finished product will have even edges and a more uniform thickness. Of course you can also use a regular round pan, but you will need to trim the omelette after it is folded up. Try to ensure that the egg mixture does not fry or turn brown or that the liquid burns off. The layers must be compact and should not separate. While it is still warm, the finished omelette is pressed into shape with a bamboo rolling mat, which will also imprint a nice surface texture on it.

You can achieve an extra, delicate effect by placing a sheet of *nori* in the omelette and incorporating it in the folds. When the omelette is sliced, the *nori* will stand out as fine black folded strands against the yellow of the omelette.

OMELETTE (*TAMAGO-YAKI*)

Omelette is excellent both as a *nigiri*-zushi topping and as *maki*-zushi filling. The yellow colour from the egg yolk provides an interesting contrast to red and white fish and the green tones from avocado and cucumber.

Once prepared, the omelette is cut into thin slices, which are placed on the rice ball and fastened with a strip of *nori*.

Tamago-yaki for *chirashi*-zushi is usually cut into somewhat shorter, thicker pieces, sometimes with a pointed end that is placed vertically for decorative purposes.

Folded omelette (*tamago-yaki*) for sushi.

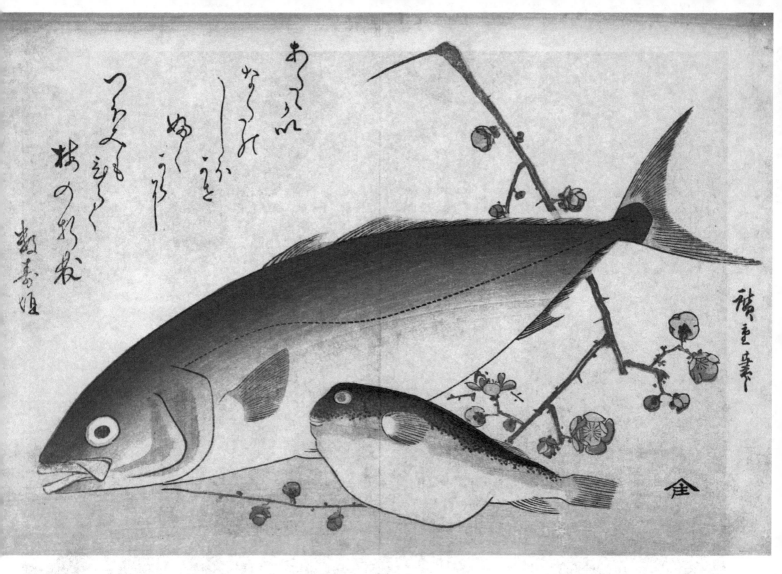

SUSHI WITH A DIFFERENCE

Every culture has its own eating habits. What some regard as ordinary, others will often consider to be peculiar and not especially palatable. The Japanese say that everything that is fresh and that can be eaten raw, should be eaten raw, for example, as sushi and *sashimi*. This also goes for some strange things, such as the world's most poisonous fish, sea urchin roe, fish sperm, whole crab, and raw horsemeat.

Fugu– A POISONOUS PLEASURE

No book about sushi is complete without the story of the *fugu* (pufferfish or blowfish), if only because it is such an oddity and eating it is a flirtation with death. The fish is rarely available outside of Japan, where it is caught in the waters near Osaka and around Kyushu. *Fugu* is a small and rather ugly spherical-shaped fish; some species have quills like a porcupine and are called porcupine fish. By swallowing water and inflating its stomach, the fish can blow itself up so that it looks like a much larger ball. The liver and ovaries of *fugu* contain a deadly nerve poison, *tetrodotoxin*.

The poison, which cannot be rendered harmless by cooking, is a thousand times more potent than potassium cyanide. It works by paralyzing the muscles and the respiratory organs. The victim remains fully conscious and dies of asphyxiation. There is no known antidote.

Yosa Buson (1716-1783), the renouned *haiku* poet, immortalized the *fugu* in the following poem:

> I cannot see her tonight
> I have to give her up
> So I will eat *fugu*

Fugu is eaten raw as *sashimi* and cooked in soups.

The taste of the fish is not particularly interesting, but the fish is famous because it is rare and expensive, and a certain thrill is associated with eating it. If the poison is not removed absolutely correctly and seeps into the fillet, it can have fatal consequences. Sushi chefs must obtain a special certification to be permitted to prepare and serve *fugu*.

While it is supposedly not very difficult to remove the poisonous parts of the fish, the artistry consists in allowing a miniscule amount of the toxin to seep into the muscle when the fish is being filleted. The poison consumed from *fugu*-zushi or *sashimi* results in a pleasing numbness of the mouth and tongue. I have personally experienced this sensation while eating *fugu* and I have to confess that it was sort of exciting.

学術 **DEADLY POISONOUS.** *Tetrodotoxin* is named for the porcupine fish, *Tetraodontiformes* which, as indicated by the name, has four large teeth. The poison acts by blocking the sodium channels in the *membranes* of the nerve cells, causing the electrical impulses of the nerves to shut down. Less than one milligram, which is to say the amount that could be placed on the tip of a needle, is sufficient to kill an adult. The *fugu* itself does not secrete the poison; it is derived from the bacteria which live in the fish. It has a mutation in its own sodium channels that renders it immune to the effects of the poison.

珍談 **IT IS SAID THAT** a couple of hundred people die every year in Japan after eating *fugu*. It is also said that these fatalities are not due to sushi eaten at a sushi bar where it can be served only if the chef has a special licence to prepare *fugu*.

Fresh sea urchin roe (*uni*)
on a wooden tray.

SEA URCHIN ROE (*UNI*)

As far as I am concerned, there is no doubt that *uni* is one of the most sublime ingredients for sushi. *Uni* is sea urchin roe, in fact, it is actually the whole reproductive organ. It is eaten raw, usually as battleship sushi.

The taste of *uni* is a concentrated blend of fresh salt water, iodine, and *bromophenol*, which we normally associate with the pleasant tang of a brisk sea breeze. The taste of *uni* is attributable to its diet of seaweed and algae. It is slightly creamy due to the significant fat content and there are nut-like nuances as well. I particularly enjoy *uni* at the conclusion of a meal of sushi, because while eating the other varieties I am filled with the joy of anticipation of this grand finale.

Uni is not widely available where seafood is sold, as sea urchins are not harvested for retail sale in many places. In southern Europe, sea urchin is used to add flavour to soups and fish soufflé. Sometimes *uni* is sold canned as a sort of paste, but the taste of this product is regrettably much inferior to that of real fresh *uni*.

調
理
法 *Uni.* The shell of the freshly caught sea urchin is cut open and the exposed roe is removed and cleaned carefully with water to which a bit of pickling salt has been added. The reproductive organs of both male and female sea urchins are usable, but *uni* from females is sweeter and considered more desirable.

Uni in battleship sushi.

LONG NECK CLAM OR GEODUCK (*MIRUGAI*)

Even though the long neck clam (*Panope generosa*) is found only in the waters around Japan and along the Pacific Coast of North America, I have included it because it is one of my absolute favourites for sushi. It can grow to a very old age and great size, weighing several kilograms, and lies buried in the seabed, often a metre below the surface. The clam is connected to the water by a 'trunk' which is up to a metre in length. Through it, the long neck clam gets its food supply and expels its waste products.

Mirugai is eaten as *nigiri*-zushi or as *sashimi*. It is probably the strongest tasting of the shellfish used for sushi, so it can easily do with a little extra *wasabi*.

The 'trunk' is almost crunchy crisp and gristly to chew. This consistency, as well as the taste of salt water that is released when one bites into the clam, produce a very unusual sensation, which makes *mirugai* into a taste experience like no other.

Mirugai nigiri-zushi is prepared from the long 'trunk' of the geoduck.

調理法 **MIRUGAI**. The 'trunk' is scalded and the outer part is cut away. The underside is sliced into pieces of appropriate size, which are pounded lightly with the end of the sushi knife to tenderize them. The edge of the piece facing the outer tougher side can be made easier to chew by making a few nicks in it. This will also facilitate fastening the piece on the rice ball when making *nigiri*-zushi.

Long neck clam or geoduck (*mirugai*). The large 'trunk' is used for sushi.

Amaebi-zushi, two sweet raw shrimp as *nigiri*-zushi.

珍
談
It is said to be a most delightful and sought-after pleasure to eat what are called 'dancing shrimp' (*odori*), which essentially are live shrimp that are eaten so quickly after the peel is removed that they move in the mouth when one eats them. *Odori* are not considered part of a traditional sushi meal.

The end of a spider roll made from crab (*kani*) that has just moulted as part of its growing process (soft shell crab).

Sweet shrimp (*amaebi*)

If they are to be eaten raw, shrimp must be absolutely fresh. These raw shrimp are referred to as sweet shrimp, *amaebi*, because they have a significant content of free *amino acids*. They are gelly-like and translucent. In comparison with cooked shrimp, which are firmer and stiff, they have a soft, creamy texture. *Amaebi* can be purchased shelled and frozen.

The shells are left on the tail ends and often two smaller shrimp are placed side by side on one rice ball for *nigiri*-zushi and are held in place by a narrow strip of *nori*. As raw shrimp are delicate and break easily, they must be handled with great care.

Amaebi are eaten as *nigiri*-zushi and *sashimi*.

Soft shell crab (*kani*)

Normally one eats only the muscle meat in the legs and claws of crabs. The exception is soft shell crab, which is deep fried and eaten whole. Soft shell crab is not a separate species of crab, but a crab that has moulted as part of the growing process. As the crab's skeleton is on the outside of its body, it cannot grow without at the same time changing its shell. After the old shell has been discarded and before the new one has had a chance to harden, the animal is soft and unprotected and, consequently, easy to eat. In these first few hours after the moult, the soft shell crab is considered a delicacy for making such things as *maki*-zushi or *temaki*-zushi. A *maki* roll with the legs of the crab sticking out of one end has a slight resemblance to a spider. That is why this type of roll is called a spider roll in North America, but it can be hard to find in other parts of the world.

One usually encounters crab only as a cooked filling in *maki*-zushi, sometimes even in the guise of the synthetic crabmeat, *oboro*, also known as *denbu*. *Oboro* is made from minced, cooked fish that has been set with gelatine and pressed into shape, sometimes as a long stick which has been dyed red on one side to resemble the muscle meat on the leg of a large crab. *Oboro* is also used by some sushi chefs as topping in *chirashi*-zushi.

FISH SPERM (*SHIRAKO*)

The small sacs containing fish sperm (milt) are considered a delicacy in Japan, especially those from *fugu*. The sperm is white and viscous and is eaten lightly salted. It is only available during winter.

Shirako is eaten raw, grilled, or in soup. It can also be served as battleship sushi.

Shirako does not look very appetizing and I had to convince myself to eat it when one of my Japanese colleagues, Aki Kusumi, persuaded me to try it on a visit to Nagoya. Even though I must admit that it tastes quite good, the sensation of having this rather sloppy bit of food flopping around in my mouth was rather strange.

RAW HORSE (*UMA*)

If one can eat raw beef as carpaccio and raw fish as *sashimi*, why should one not eat raw horse flesh, *uma*? *Uma* is also called *sakura niku*, which literally translated means 'cherry coloured meat'. Preferably *uma* should be lightly marbled. It is sliced very cold from the freezer and is eaten immediately so that the taste of the fat does not become overpowering.

Sliced into thin pieces like *sashimi*, *uma* is served under the name *basashi*.

My Japanese colleague, Motomu Tanaka, was kind enough to introduce me to *uma* on a visit to Kyoto. It was served to us almost directly out of the freezer, sliced into fine, thin slivers and dipped in a little soy sauce. The taste was interesting and sweetish, but I am not likely to undertake a long journey to eat this dish again.

THE REST OF THE MENU

The moment two bubbles are united
they both vanish
a lotus blooms
Kijo Murakami (1865-1938)

SIDE DISHES AND CONDIMENTS

In a sushi restaurant or a sushi bar it is not unusual to order a mixture of sushi and other types of food. First and most prominent among these is raw fish without rice, *sashimi*, as well as small servings of hot food, vegetables, and condiments. The possibilities for combining these dishes with sushi are limited only by the imagination and inclination of the individual. Similarly, Japanese inspired culinary creations are often incorporated into conventional Western meals as appetizers, side dishes, or main courses.

SASHIMI – JUST PLAIN RAW

Sashimi is easy to prepare because it generally consists of only sliced raw fish and shellfish. With a fresh fish at hand, you can make *sashimi* in a matter of minutes. It sometimes occurs to good friends of mine who go fishing to slice up fish that they have just caught, right in the boat, and eat them on the spot. This is probably exactly what our distant ancestors did in prehistoric times.

All fish and shellfish which can be eaten raw can be made into *sashimi*. Given that *sashimi* by definition indicates something that is cut up, oysters cannot, strictly speaking, be classified as *sashimi*. Also, there are fish and shellfish which cannot be consumed raw, but which must first be cooked, salted, or marinated. Cases in point include shrimp, cuttlefish, and mackerel. Slices cut from these seafoods can also be used for *sashimi*.

Preparation of *sashimi*, however, amounts to more than just cutting up pieces of fish and, as in many other things in life, the devil is in the details. The precise way that various types of fish and shellfish are sliced, the combination of ingredients, and especially the presentation are all elements that elevate *sashimi* to the level of art. A professionally arranged platter with *sashimi* is a feast for the eyes. Often the fish is presented on bamboo leaves together with thin strips of fresh radish (*daikon*) and small ice cubes. The interplay of colours can be enhanced by a careful choice of decorations – different types of roe, a pair of green *shiso* leaves, or a small fan of finely sliced avocado. Fish that is unskinned, for example, shiny mackerel, adds a special dimension.

Red tuna (*maguro*) *sashimi.*

Soft fish, such as salmon and tuna, are cut into thicker slices than firm fish, such as flatfish and ocean perch, or octopuses.

Ikizukuri is a particular, slightly bizarre type of *sashimi*, which some might consider rather off-putting. To make it, a fish that has just been killed is cut up and artistically reassembled on the skeleton before being served.

Sashimi can also be prepared as *tataki*. The fillet of the fish is first lightly seared on all sides and then sliced. *Tataki* is especially impressive if made with red tuna because the deep red of the raw fish really stands out against the cooked brown edges where the *myoglobin* of the muscles has lost its colour.

Often *sashimi* makes up the first part of a Japanese meal; it is typically served before the sushi. It is eaten by dipping the individual pieces in soy sauce into which *wasabi* has been mixed. Between bites, the palate is refreshed with a little picked ginger, *gari*.

Red tuna (*maguro*) *sashimi* prepared as *tataki*, by searing it lightly.

Pickled radish (*daikon*), *takuan-zuke*.

TSUKEMONO – PICKLES

A variety of pickled vegetables, *tsukemono*, make an appearance as condiments at all Japanese meals, both hot and cold. In connection with sushi and *sashimi*, pickled ginger (*gari*) is first and foremost among them.

The pickled vegetables are eaten, in no particular order, in between the other dishes. *Tsukemono* offer an appetizing array of flavours that reflect the sour, salty, herbal, and spicy elements that went into their preparation. *Nuka-doko*, the fermentation medium based on rice bran, imparts an especially rich taste. Among its other uses, it is utilized for making *takuan-zuke* from white radishes (*daikon*).

For me, there is a distinctive taste experience linked to the crunchy mouthfeel that is bound up with the best types of *tsukemono*.

Tsukemono made from different vegetables, some of which are dyed with red *shiso*, come in a veritable riot of colours and shapes.

Tsukemono of pickled eggplants (top right) and cucumbers (the other three).

EDAMAME – GREEN SOYBEANS

Soybeans are the most fundamental vegetable foodstuff that one can think of – a very large proportion of the combined global protein production for human nutrition is derived from soybeans.

Very young, small green soybeans are a delicacy. They are sometimes available fresh in Asian food stores, but in some places frozen ones of an acceptable quality can be found more readily.

Edamame are eaten as a snack or as a side dish. The cooked beans are served in a bowl or on a plate and are eaten with the fingers. Only the bean itself is eaten – put the shell in your mouth and then gently push out the beans. Usually there are one to three of them in an *edamame* pod. They taste best when they are still slightly warm, but they can easily be eaten in between the various dishes during the meal, even if they have grown a little cold.

This might all sound somewhat unsophisticated – but just try them; they are delicious!

調理法 *Edamame* – green soybeans. Immerse the frozen *edamame* in lots of boiling water and cook them for only a brief time, one minute at most. Fresh beans should be boiled just a little longer, for two to three minutes. They should be cooked through but should still be crisp. Drain the beans and rinse them quickly in cold, running water. Before the beans have a chance to dry completely, toss them with a little coarse salt.

Edamame – cooked, small green soybeans sprinkled with a little coarse salt.

Dried and toasted red seaweed, *dulse*.

調
理
法 **TOASTED SEAWEED.** The seaweed fronds are placed in a toaster oven and toasted for a few minutes until they are crisp. Note that the pieces of seaweed are soft while still warm just after toasting, but they will quickly become crisp as they cool.

学
術 **SALT AND HIGH BLOOD PRESSURE.** Preferably one should avoid ingesting excessive quantities of ordinary cooking salt, sodium chloride. People with high blood pressure have to be particularly careful about having too much salt in their food. In these cases, ground seaweed can be a good substitute for cooking salt. It often contains a surplus of potassium salts compared to sodium salts, which impart a similar taste, but have a favourable effect on blood pressure.

TOASTED SEAWEED

Some types of seaweed make a wonderful snack before a sushi meal. Only seaweed with delicate and thin fronds can be used. *Konbu*, kelp, is generally too thick and it can be quite tough, so it is normally cooked before eating.

For toasting, I have become especially fond of two types which are harvested in the western part of Canada, on the Pacific side of Vancouver Island. They are macrokelp (*Macrocystis integrifolia*) and, best of all, bull kelp (*Nereocystis luetkeana*). The fronds of these two species are thin and delicate.

One should never be concerned about the white spots which can appear on toasted seaweed. They are caused by deposits of salt and MSG, *monosodium glutamate*, which impart the sweet *umami* taste. MSG helps to give the seaweed its characteristic taste, which makes it delectable both as a snack and in soups.

A red seaweed, known as dulse (*Palmaria palmata*), is native to the North Atlantic and North Pacific. It is commonly harvested in Brittany, Iceland, Ireland, Maine, and Atlantic Canada. It, too, makes an excellent, nutritious snack when toasted. An especially delicious variety is the applewood smoked dulse that comes from Maine.

Freshly toasted bull kelp or dulse can be crumbled and sprinkled on a salad, on marinated fish, or on a bowl of rice.

Bull kelp, dried and toasted.

Fu – 'THE MUSCLE OF THE DOUGH'

Fu, or *seitan*, is made from wheat, which has a considerable content of *gluten proteins*, between 8 and 18%. A dough is made by kneading together wheat flour and water, in which the gluten cannot dissolve. This dough is then repeatedly rinsed in water to wash away the starches and fats, leaving a firm, elastic mass. This mass has a very significant protein content, typically 30%, and only a little fat, about 2%. The taste of this protein mass can be intensified by simmering it in soy sauce, to which a little ginger or garlic has been added.

The Chinese name for *fu* is 'mien chin', which means 'the muscle of the dough'. Mien chin was already used by Buddhist monks in the 11th Century as a meat substitute. It is said that some vegetarians avoid *fu* because its taste and consistency are all too reminiscent of meat. *Fu* can also be *fermented*. As gluten proteins contain a significant amount of the amino acid *glutamic acid*, fermentation causes the formation of MSG, *monosodium glutamate*, which has the *umami* taste.

When combined with other ingredients, for example, avocado and seaweed (*wakame* or *nori*), *fu* can be served as part of a sushi meal. The firm, elastic consistency of the gluten mass nicely complements the soft consistency of avocado. Well-made *fu* makes a pleasant squeaky sound when it is chewed.

In Japan, *fu* is also known by the name *nama fu* and incorporated into the classical vegetarian temple food, *shōjin ryōri*. *Fu* is used in soups or broiled and roasted, and can be be served in combination with mushrooms and hot vegetable dishes.

Fu or *seitan* – gluten mass with a very high protein content.

調理法 *Fu* WITH AVOCADO. Any remaining liquid is drained from the *fu* and it is sliced into suitably sized pieces which are placed on a bed of avocado. Soy sauce is drizzled over it and toasted, crumbled seaweed or *furikake* is sprinkled on top just before it is served.

学術 THE WORD GLUTEN designates a series of *proteins* found in wheat. These are very long molecules, especially *glutenin* and *gliadin* which contain up to a thousand *amino acids*. Gluten proteins are not water soluble and when they are mixed with water they bind themselves together to form very long chains. These chains make the mixture elastic and tough. We know these properties from dough kneaded for baking bread. The elastic structure makes it possible to gather and hold on to small bubbles of carbon dioxide formed by the yeast or baking powder used as a leavening agent. The gluten content of a particular cereal grain, therefore, determines how suitable flour made from it is for baking. There is no gluten in rice and, consequently, it is rarely used for baking.

Fu with avocado, toasted white and black sesame seeds, and a little finely cut up *nori*.

Shiitake. The mushroom is named after the tree on which it grows, *shii*, and the word for mushroom, *take.* You can grow your own supply if you procure a hardwood log, possibly oak or beech, which has been inoculated with *shiitake mycelia.*

SHIITAKE

After the common button mushroom, the *shiitake* (*Lentinus edodes*) is the most widely cultivated mushroom in the world. It grows naturally on the logs of the evergreen *Castanopsis cuspidata* tree (a relative of the beech and oak, known as *shii* in Japanese) and has been cultivated in China since the 1100's. In recent decades, *shiitake* cultivation has become intensive and commercialized, spreading to the rest of the world, especially the West. It is one of the most widely utilized mushrooms in Chinese and Japanese cuisine.

Fresh *shiitake* keep well in the refrigerator for up to a month and can be eaten as is. They have a woody taste and are a little acidic. One reason for their popularity is that they are very easily preserved in dehydrated form and have virtually unlimited shelf life if left in an unopened package.

Dried, finely sliced *shiitake.*

The aesthetic quality of the dried *shiitake* is often based on its having a perfect mushroom cap shape and on the beautiful patterns formed on it. The price of dried Japanese *shiitake* fluctuates greatly, depending on the appearance and the varietal in question. The most highly prized type is *donko* which has a small, dark cap with a flowerlike design.

Drying concentrates the flavour of the mushroom, this being especially due to the formation of a chemical substance, *lenthionine.*

Like other mushrooms, *shiitake* has a distinct *umami* taste, due to the presence of free amino acids, especially *glutamic acid.*

調
理
法
MARINATED *SHIITAKE* mushrooms. Wash dehydrated *shiitake* in cold water and then soak them for two hours, or so, in lukewarm water. Drain the water and reserve it for use in soups. Remove the tough stems and slice the remaining caps into strips, if you so wish. Place the mushroom pieces in a marinade consisting of soy sauce, sweet sake (*mirin*), or a little sweet white wine. Let them marinate for a few hours and adjust the saltiness with the help of soy sauce. Once in the marinade, the mushrooms can easily stand overnight in a cold place. When you serve them put a little of the marinade aside to use as a dipping sauce. If the *shiitake* are to be used in *futomaki* or pressed sushi (*oshi*-zushi), squeeze most of the liquid out of the mushrooms before proceeding further.

Shiitake contain very little *fat*, but almost as much *protein* as soybeans. They are rich in minerals and *vitamins*, among them vitamin B and, more particularly, vitamin B_{12} which is not found in vegetables. Research has indicated that *lentinan*, a substance found in *shiitake* which helps to boost the immune system, may be effective as an anti-cancer agent.

As *shiitake* have a very strong taste, they are used only sparingly in conjunction with a sushi meal. I have three personal favourite recipes based on marinated *shiitake* mushrooms: as filling in thick *futomaki* rolls, as a layer in pressed sushi, *oshi*-zushi, or simply as a small mushroom salad side dish.

MUSHROOMS, THEIR AROMA, AND CANCER

The aroma of freshly harvested mushrooms is due to a short-chain alcohol, *octenol*, which is formed by the *enzymatic* breakdown of the polyunsaturated *fatty acid, linoleic acid*. This process is intrinsic to the defence system of the mushroom, which is set in motion when its cells are damaged in any way, especially those of the lamellae, the gills on the underside of the cap. Immature mushrooms, therefore, have a milder taste than those with fully developed lamellae. Also, brown mushrooms are more flavourful than white ones.

The special aroma of dried *shiitake* can be attributed to the *enzymatically* mediated production of *lenthionine*. This substance is named after the mushroom's Latin botanical name, *Lentinus edodes*. *Lenthionine* is a ring-shaped organic molecule which contains carbon and sulphur. Large quantities of *lenthionine* are formed when the *shiitake* is dried and when it is soaked in tepid water. Quick cooking or frying ruins the enzymes and this results in less *lenthionine*.

There is scientific evidence that certain *polysaccharides*, such as *lentinan*, in the *cell walls* of *shiitake* act as anti-cancer agents. In addition, the mushroom contains substances, including *lenthionine*, which possibly suppress the formation of cancer-causing *nitrosamines* in the digestive system.

Marinated *shiitake* mushrooms.

Avocado with *wasabi* and *gari*.

学術 **FRUIT OR VEGETABLE?** In the botanical sense, avocado is a fruit because it has a stone from which a new plant can sprout. In the gastronomical sense, avocado is thought of as a vegetable.

調理法 **AVOCADO WITH *WASABI*.** Coat pieces of avocado gently with a thin sauce made from *wasabi* and lemon juice or a bit of *gari* marinade. Place them in a bowl or arrange them on a plate and sprinkle with crumbled, dried seaweed or *furikake*.

調理法 **AVOCADO WITH *GARI*.** This is one of my favourites. Cut up ripe, firm avocados into slices or cubes and mix them gently with a small amount of *gari* that has been chopped finely or sliced into thin strips. Include a bit of the marinade from the *gari* to prevent the avocados from turning brown.

This recipe can be combined with the one above by adding a little *wasabi*.

AVOCADO

There are many species of avocado, which is a fruit belonging to the laurel family. Avocados contain very little *sugar* or *starch*, but have a very high proportion of *fats*, typically 15-30%. Monounsaturated *fatty acids* predominate, of which about 70% is *oleic acid* as in olive oil, and there is also a significant presence of *omega-6 fatty acids*, with about an 11% *linoleic acid* content.

The sliced surfaces of an avocado quickly start to turn brown after it is cut up because of *oxidation* and *enzymatic* activity. This can be prevented by brushing the surfaces with something acidic, for example, lemon juice, to inhibit the enzymatic activity. It can also be inhibited by lowering the temperature or shielding the surfaces from contact with air. The latter is most easily done by covering them with plastic wrap or placing the avacado pieces in water, preferably with a touch of lemon juice.

RIPENING AND BROWNING OF AVOCADOS

The best way to ripen avocados is to leave them at room temperature until they reach the right degree of ripeness. Placing them in the refrigerator slows down this process. On the other hand, it can be speeded up by placing the avocados in a closed paper bag, which will concentrate the natural ripening gas, *ethylene*, released by the fruit itself. It is important to use a paper bag, as opposed to a plastic bag, as it is porous and will permit the exchange of oxygen and carbon dioxide, both needed by the fruit as it is still alive. One can also further accelerate the ripening process by putting a ripe tomato or a banana in the bag with the avocados. These two fruits, when ripe, give off great quantities of ethylene.

Both fruits and vegetables take on a brown discolouration when they are sliced, mashed, or otherwise worked on mechanically. The browning is due to an *enzymatically* controlled *oxidation* of the *phenol* compounds in the plant. This causes molecules which absorb light to aggregate and results in the brown colour on the surface. It can be retarded by brushing the cut surface with lemon juice, which helps to denature the enzymes. The browning effect is part of the plant's natural defence system, as it makes the fruit look less appealing. The active enzymes lie hidden in the cells' *vacuoles*, small cavities inside the cells. They are released from the vacuoles when an insect or a microorganism attacks the plant. The enzymes are meant to harm the attacker.

調
理
法
AVOCADO WRAPPED IN *NORI*. Peel and slice an avocado lenghtwise. Brush the slices lightly with a little marinade from *gari* to prevent browning. Wrap each slice in a piece of *nori* just before serving. The wrapped avocado slices are eaten with the fingers and dipped in soy sauce, to which a little *wasabi* can be added.

Different species of avocado have a somewhat individual taste. A good selection from different countries is widely available, with the type largely dependent on the season. I prefer the ones that are not too watery, with my personal favourite being the small Californian variety called Hass, which has a creamy, nutty flavour. The ripe Hass avocado has a very dark green granulated peel and a small stone. Its taste is fuller than that of the slightly larger Fuerte avocado, recognizable by its smooth, thin pale green skin.

As avocados have a mild aroma and taste attributable to a mixed group of chemical substances (*terpenes* and short-chained *fatty acids*), they are a natural and versatile complement for raw fish. For a sushi meal, avocados can be used as a side dish, in a *maki* roll, as a bed on which *sashimi* is placed, as a green decoration with both red and white fish, or simply wrapped in a piece of *nori*.

Eggplants.

OVEN DRIED EGGPLANTS WITH RED *SHISO*

Eggplants that are baked or dehydrated take on a dry, slightly spicy taste, which reminds me of the smell of a forest in autumn. The distinctive taste of red *shiso* (*aka-shiso*) complements them surprisingly well and its delicate aroma is brought out especially strongly if the eggplants are a little warm and moist when the *shiso* is added to them.

Oven dried eggplants with red *shiso* make an excellent side dish for sushi or *sashimi*.

調
理
法
OVEN DRIED EGGPLANTS with red *shiso* (*aka-shiso*). Wash eggplants which are not too large in cold water and cut off the tops. Unless the eggplants are very small, slice them lengthwise and then cut the halves into slices that are about 5 mm (¼ inch) thick. Place the slices on baking paper on a roasting pan and bake in a 200-250°C (400-475°F) oven for about 30 minutes. Turn the slices once in a while so that they will dry evenly. They are ready when they feel dry to the touch and are slightly brown, but are still a little moist on the inside. Remove them from the oven and sprinkle them with red *shiso* (*aka-shiso*), either in the form of crushed whole dried *shiso* leaves or prepared *yukari furikake*.

Oven dried eggplants with dried, red *shiso*.

ZUCCHINI

Zucchini, sometimes called squash or courgette, belongs to the marrow family. Small, very fresh zucchini can be eaten raw in a salad, possibly together with their delicate, showy, bright yellow flowers. This vegetable can be used in many ways and can easily be slipped into a Japanese inspired meal.

Small zucchini.

Sautéed or steamed zucchini are one of my favourite accompaniments for fish and sushi. Warm zucchini on their own have a slightly boring, flat taste. But combined with the flavours of *gari* and toasted bull kelp or *nori* they are superb.

Notice the aroma of the dry seaweed when it becomes moist and warm after being sprinkled on the still hot pieces of zucchini.

調理法 SAUTÉED ZUCCHINI *SENGIRI*. Select zucchini which are not too large. Wash them and cut them in julienne strips, which in Japanese are called *sengiri*. Sauté finely chopped *gari* for a short time in a pan with a little grapeseed oil (recommended because it has practically no taste of its own). Add the strips of zucchini with the heat on high. Turn them frequently with a spoon and do not let them brown. If the pot is covered between stirrings, the zucchini will be steamed due to the liquid they release when heated. But be careful that they do not become too soft; they should retain a little crispness. Turn the zucchini strips out into a bowl and sprinkle them with crushed, toasted seaweed (*furikake*). Toasted, crushed sesame seeds can also be added as an additional topping. The dish must be served immediately before the zucchini becomes limp and before the characteristic smell of the warmed up seaweed fades away.

If you think that the dish should have a more definite taste, you can sauté a little coarsely chopped garlic with the *gari*.

Sautéed zucchini with *furikake*.

Common purslane.

Miner's lettuce.

Glasswort stem.

THE PORTULACA, OR PURSLANE, FAMILY

Two members of the portulaca, or purslane, family are eminently suitable for use with a sushi meal. Although they form no part of traditional Japanese cuisine, I am including them just the same because the green leaves of the plant are splendidly decorative on an arrangement of sushi. When paired with avocado they make an excellent side dish. Unfortunately, these plants are shamefully overlooked in the contemporary kitchen.

Common purslane (*Portulaca oleracea*) is sometimes called pigweed or little hogweed, an indication that in the past it has been scorned as food for humans. It is a small, succulent herb, which has thick leaves and slightly rubbery, but crisp, stems. The plant is rich in calcium and the superunsaturated *omega-3 fatty acid, alpha-linolenic acid*. In contrast to other decorative green herbs, such as watercress, which often have too harsh a taste, the delicate, lettuce-like taste of purslane is a brilliant accompaniment to sushi.

Winter purslane or miner's lettuce (*Montia perfoliata*) is a more slender plant, which has cup-shaped leaves growing as a disc around the stem. It is an important source of *vitamin* C. Its flavour also combines well with sushi and the stems, which are often quite long, make an attractive little interlaced clump of green as decoration on a plate of *sashimi*. Toward the end of the growing season, the plant has small white flowers which are also edible.

GLASSWORT

Glasswort (*Salicornia*) is also called sea asparagus, pickleweed, and marsh samphire. It is a succulent annual which has fleshy green jointed stems with small scale-like leaves. Salt water tolerant, glasswort grows in marshy areas. It is not yet sold widely commercially and is highly seasonal, but you may occasionally find it in a fish store or as a frozen product.

Only the newest, youngest shoots of glasswort are worth eating. In bygone days, it was cooked or preserved and eaten as a sort of substitute for asparagus. I prefer to utilize it as a side dish for sushi or as decoration on an arrangement of *sashimi*. The salty taste is a pleasant complement for fresh fish and shellfish, in particular oysters.

調
理
法
Marinated mackerel with toasted seaweed. Slice the marinated mackerel fillets on the diagonal. The smaller and thinner the fillet, the more it should be sliced on the diagonal. The pieces should be about 5-7 millimetres (¼ inch) thick. Drizzle a little soy sauce over the fish and sprinkle dried, toasted seaweed on top just before serving.

Marinated mackerel with toasted seaweed

This is a dish which is easy to prepare on short notice if you have marinated mackerel in the freezer. A fat, frozen mackerel can be cut into slices about 5 to 10 minutes after it is taken from the freezer and the dish can be arranged before it has defrosted completely. It makes an exceptionally easy to prepare appetizer.

The dish can be served in individual portions or presented on a platter or in a bowl.

Marinated mackerel with soy sauce and toasted seaweed.

SHŌYU MARINATED SALMON

If you prefer not to eat salmon completely raw, you might enjoy this recipe as it preserves the good fats to be found in the fish.

Normally it is difficult to grill or fry salmon in smaller pieces because the fish breaks apart easily, even after a short cooking time. However, marinating the salmon in soy sauce denatures the *proteins* in the outer layer and makes the fillet firmer. Doing this first prevents the fish from falling apart so readily in the frying pan.

As the salmon pieces are seared very lightly, the inner part of the fish stays very red and almost raw.

The resulting appealing texture – firm on the outside and soft and juicy on the inside – makes marinated salmon an interesting side dish for sushi or a fit accompaniment for cooked rice, possibly served with avocado or oven roasted eggplants on the side.

Soy sauce marinated salmon can also be eaten as is without any cooking, just like *sashimi*.

調理法 *SHŌYU* MARINATED SALMON. Slice fresh salmon into pieces that are about 2 centimetres (¾ inch) thick and marinate them in soy sauce to which a little *wasabi* has been added. Leave them in the marinade for 5 minutes to half an hour – the longer time will result in a stronger taste. Place the marinated salmon pieces on a not too hot ungreased frying pan and sear them quickly on each side. Serve them as is or with warm rice.

Shōyu marinated salmon.

調
理
法
OVEN BAKED SALMON with *gari*. Place pieces of salmon, cut crosswise from the fish into slices about 3-5 centimetres (1½ inches) thick, in an oven-proof dish or on baking paper in a roasting pan. They can be baked with or without the skin, but they will have a more intense flavour if the skin is left on, facing downward. Sprinkle finely chopped *gari* on them and bake in a 200°C (400°F) oven for 5-10 minutes, depending on their size and on how rare they should be. I prefer mine very rare and I find that a shorter cooking time leaves them more juicy. Serve the salmon with rice, vegetables, oven dried eggplants, *fu*, or avocado with *gari* and *wasabi*.

OVEN BAKED SALMON WITH *GARI*

This dish is also easy to prepare and everything can be made ready and put in a warm oven for 10-15 minutes before the dish is to be served.

Gari accentuates the mild taste of the salmon, precisely as it does in combination with sushi or *sashimi*.

Oven baked salmon with *gari*.

Fresh *enokitake*.

ENOKITAKE

Enoki mushrooms (*Flammulina velutipes*) are also known as *yukinoshita*, which means 'under the snow'. The mushrooms grow in a little tight bunch of long, white stems with small ball-shaped heads.

Enokitake have a fresh mushroom taste and are very crisp. They can be eaten raw in salads or in the traditional Japanese dish *shabu-shabu*, where they are quickly cooked in a *dashi* broth. For sushi meals I use *enokitake* on their own for artistic purposes, for example, as a decorative flourish at the end of a *maki* roll.

SUSHI RICE BALLS

This side dish is based on one of the earliest elements of Southeast Asian cuisine: simple, hand shaped sushi rice balls, which are eaten with the fingers, with or without first being dipped in soy sauce. To spice up the taste of the rice, you can sprinkle them with green *shiso* leaves, *furikake*, toasted sesame seeds, or finely chopped *umeboshi* plums. Another possibility is to wrap a salted green *shiso* leaf around the rice ball.

As sushi rice balls can be eaten either warm or cold, they are super in a bag lunch. They can also be turned into *onigiri* by wrapping them in a piece of *nori*. Because it absorbs moisture, the seaweed should be wrapped around the rice ball only just before it is to be eaten.

調理法 **RICE BALLS WRAPPED IN *SHISO*** leaves. Soak salted, green *shiso* leaves (*ao-jiso*) in water, dry them, and lay them out on individual pieces of plastic wrap. Place a rice ball of suitable size on each leaf and then wrap the *shiso* tightly around the rice ball by bringing the corners of the plastic wrap together and giving them a twist. Remove the wrap carefully and serve the rice balls with the leaf side up.

Rice balls (*temari*) wrapped in salted green *shiso* leaves.

Rice balls with thinly sliced green *shiso*.

調理法 **RICE BALLS WITH GREEN *SHISO*.** Hand shape balls about 3-4 cm (1½ inches) in diameter from sushi rice. Cut green *shiso* leaves into very thin strips and sprinkle them on top. The rice balls can be served either warm or cold arranged in a row on a narrow platter. If the rice balls are very warm, the *shiso* leaves should be sprinkled on just as they are to be eaten, otherwise the fresh *shiso* will turn brown and look unattractive.

SOUPS AND SALADS

Japanese cuisine boasts an abundance of several different soups which are incorporated into virtually every meal, including breakfast. Japanese soups are very simple and can be prepared quickly with only a few ingredients. Clear broth (*dashi*) and *miso* soup are the mainstays and often accompany a sushi meal. Small salad dishes are also excellent with sushi, especially when they make use of different types of seaweed.

DASHI – FISH STOCK

Most Japanese soups are based on a fish stock, *dashi*, which is made from dried fish flakes, *katsuobushi*, cooked together with seaweed, *konbu*. The *konbu*'s contribution to the stock are the soluble minerals and *amino acids* that produce the *umami* taste. As these substances are often found as precipitates on the surface of the *konbu* leaves, the seaweed should not be rinsed before use. *Dashi* is the foundation for the well-known *miso* soup, as well as soups with vegetables, mushrooms, fish, shellfish, and noodles. It is also the broth in which the popular Japanese equivalent of fondue, *shabu-shabu*, is prepared.

Katsuobushi is made from *katsuo* (bonito), a medium sized fish from the mackerel family found in the Pacific. Fillets from the fish are cooked, salted, smoked, fermented, and dried, ending up as rockhard pieces that will keep almost indefinitely. The *katsuobushi* is shaved into paper thin flakes using a special wooden cutting box.

These fish flakes are initially used to produce 'first *dashi*' (*ichiban dashi*), which is the basis for a delicate clear soup, *suimono*. They are then boiled to make 'second *dashi*' (*niban dashi*) for soups such as *miso*.

For those in a hurry there is another, faster way to make *dashi*. Stores carrying Japanese food products sell *dashi* powder in small, practical packets which are just sufficient to make one bowl of fish stock.

調理法 FIRST *DASHI* (*ICHIBAN DASHI*). To make first *dashi* start by placing a couple of pieces of seaweed, *konbu*, in 1 litre (4 cups) of cold water and warm it to just before the boiling point. The water must not be allowed to boil because that would cause the seaweed to give off a bitter taste. Remove the seaweed and add 10-20 grams (2-4 tablepoons) *katsuobushi* flakes. The quantity depends on how concentrated the broth is to be. Bring the mixture to the boiling point, take it off the stove, add 1 decilitre (½ cup) of cold water, and allow it to rest for a few minutes. Strain the liquid to remove the *katsuobushi* flakes. This broth is the first *dashi*, the finer of the two. It is used for a clear soup, *suimono*, or for *shabu-shabu*.

Second *dashi* (*niban dashi*). The second *dashi* utilizes both the *konbu* and the *katsuobushi* flakes left over from making the first round. To them are added 2 litres (about 8 cups) of cold water and all is brought to the boiling point. Let the mixture simmer for about half an hour. Strain the liquid to remove the seaweed and fish flakes. The result is second *dashi*, which is more cloudy and bitter than first *dashi*. It is used for such dishes as *miso* soup.

Katsuobushi shaved into paper thin flakes is well suited for the preparation of the fish stock, *dashi*.

SUIMONO – CLEAR BROTH

Suimono, based on first *dashi*, is the *Zen* of soups, a study in simplicity. The idea is not to have as much content as possible, but only a little with colours and forms which are in perfect harmony. Two slices of *shiitake*, a couple of pieces of *konbu* or *wakame*, a single shrimp, a piece of *tofu*, a small piece of white fish with skin on – nothing more is needed.

Suimono opens the door to a whole kingdom of possible variations, which is limited only by the imagination of the chef.

Suimono is served in a bowl with a lid. The lid is removed before one drinks from the soup and is replaced again to keep the soup warm in between sips. Larger pieces of fish, seaweed, and vegetables can be picked up with chopsticks.

Classical Japanese soup bowl for serving *suimono*, clear broth, here shown with pieces of white *tofu* and a few touches of green.

Katsuobushi – fish preserved five times over

Katsuo (bonito) is a medium sized fish from the mackerel family, which the Japanese preserve using no less than five different techniques: cooking, salting, smoking, dehydrating, and *fermenting*. The fish fillet is first cooked in salt water. Then it is smoked for about two weeks until the muscle is completely dry and hard. Finally it is fermented and sun-dried several times in succession. Having undergone all these processes for several months, the fish has at last been made into *katsuobushi*. A piece of *katsuobushi* is as firm and hard as wood and has taken on a reddish brown colour.

This fivefold process of conservation brings out an extraordinary arsenal of taste and aroma substances, many of which are familiar from such products as dried, salted ham and cheese. The *enzymes* of the fish itself produce *lactic acid*, sweet *amino acids*, and *nucleotides*. Smoking contributes *phenol* compounds, and cooking, smoking, and sun-drying all add cyclic *hydrocarbons* with nitrogen and sulphur. Fermentation converts the fats in the fish muscles into chemical compounds with fruity or flowery aromas.

Before use, rock-hard, *katsuobushi* is shaved into paper thin flakes which are either sprinkled on a salad, *tofu*, or rice, or used to make *dashi*, the foundation for most Japanese soups.

Whole fillets of *katsuobushi* for sale at the market in Tokyo.

調
理
法

Miso soup. To make 1 litre of soup, use ca. 1 litre (4 cups) of *dashi* and 30 millilitres (2 tablespoons) *miso* paste. If you make your own *dashi*, you will most probably use the second *dashi* for *miso* soup. As *miso* paste must not boil, it is stirred in at the end. Other ingredients are added in accordance with their cooking time. Some mushrooms and vegetables require a fair bit of cooking, whereas pieces of fish and *tofu* merely need to be warmed through. Mix the *miso* paste with a little tepid water and add it to the soup. Stir well and bring the whole to just below the boiling point. Serve immediately.

MISO SOUP

Miso soup (*miso-shiru*) is a good source of *proteins* and an excellent nutritional supplement for rice. It is made from fish stock, *dashi*, to which *miso* paste is added. Light, dark, and red *miso* can all be used.

A variety of things can be added to *miso* soup to make it more filling. For example, seaweed such as *konbu* and *wakame*, *shiitake* mushrooms, spring onions, small pieces of fish, and shellfish are all suitable. I often use the trimmings left over from slicing fish for sushi and *sashimi*. Dried seaweed and mushrooms must be soaked before being added to the soup.

The rule for *miso* soup is the same as that for clear broth: less is more. Not too many items should be added to it, nor should they be too varied – a single item, or possibly a few, of which some float on top and break its smooth surface. Just before serving, a few rings of spring onion can be added as decoration.

Miso soup is served in, and drunk from, a bowl that has a dark exterior and usually a red or black interior. Larger solid pieces in the soup are fished out and eaten with chopsticks. The dark colour of the bowl accentuates the *miso* paste, which is suspended in the soup.

Miso soup with *tofu*, spring onion, and white fish.

THE PHYSICS OF *MISO* SOUP

Everyone who has looked down into a bowl of freshly served *miso* soup has probably wondered about the patterns and currents that move about in the cloudy liquid. When the warm soup has just been poured into the bowl, the soup is uniformly cloudy. If you let it stand undisturbed for a few minutes, you will observe that the soup becomes more cloudy in some areas than in others and that, in some places, the small particles in the soup move upward, while in others they move down. This physical phenomenon is called *convection*, and the partitioning into separate upward and downward currents are called convection cells. This effect is identical to the one that can be observed near a dusty hot radiator. The dust rises above the radiator and falls in the room away from it.

In the soup, the small particles of *miso* paste are the 'dust'. The soup is warmest at the bottom because it cools down most on the surface. The warm part of the *miso* soup is lighter than the cool part and, therefore, will rise upward. The *miso* soup near the surface is colder than that at the bottom and, as the colder soup is heavier than the warm soup, this part will start to sink. This causes the soup to separate itself into small cells with upward and downward currents. They are called Rayleigh-Bernard convection cells.

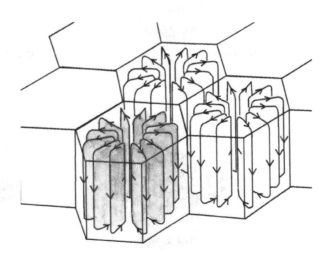

調
理
法
 SEAWEED SALAD. Soak the dried seaweed (e.g., *wakame*) in cold water for about 5 minutes and then drain the water. Dress the salad with a very small quantity of sesame oil, distributing it evenly. Crush some black or white toasted sesame seeds, or a mixture of the two, to release the delicious sesame oil from them and then sprinkle them on top as well.

Alternatively, you can toss the salad with a dressing made from wine vinegar and soy sauce, as well as a little salt and pepper.

If you prefer a seaweed salad with a bit more bite to offset the strong *umami* taste, you can carefully add a little bit of finely chopped chili.

SEAWEED SALADS

A small portion of salad made from seaweed served before a sushi meal is refreshing and awakens the appetite. Many different species of seaweed are eminently suitable and one can often purchase them as a dried assortment which is quick and easy to prepare. Alternatively, you can cut large pieces of *wakame* into smaller pieces. Note that pieces of very coarse *konbu* (kelp) should not be used in salad. They are usually too tough and often become slimy when they are soaked. This slime is made up of *polysaccharides*, which are often used as thickeners, for example, to help marmalade set. After blanching and salting, the more tender pieces of *konbu* or *wakame* (e.g., *hiyashi wakame*) can be cut into thin strips and used as a salad, possibly tossed with a little finely chopped chili.

If you wish to have a somewhat larger salad, you can mix the seaweed salad with finely cut pieces of green apple.

Seaweed salad also pairs well with a *sashimi* appetizer, for example, one made with marinated mackerel (*saba*).

Wakame salad.

Mixed seaweed salad.

Wakame salad with finely chopped chili.

Hijiki salad.

調
理
法
SIMMERED *HIJIKI* WITH CARROTS. Soak dried *hijiki* in warm water for about half an hour and then drain the water. Rince it several times in cold water. Mix the seaweed with julienned strips of carrot. Add *dashi*, a little soy sauce, and *mirin*. Allow the ingredients to simmer until most of the moisture has evaporated. Adjust the taste with sugar and soy sauce and cool the dish before serving.

SIMMERED SEAWEED

Hijiki, a brown alga, is a very nutritious food because it contains many minerals such as calcium and iodine, as well as *vitamin* B$_{12}$. While it is versatile and has many uses, it has a slightly flat taste on its own and is best combined with other ingredients and taste substances.

I think that the best use of *hijiki* is in a gently cooked dish where it is simmered in *mirin, shōyu,* and a soup stock, *dashi*. The slightly smoky flavour from the fish stock counterbalances the somewhat insipid taste of the alga. The salad is easy to make and keeps for several days in the refrigerator.

Simmered *hijiki* seaweed with carrots.

調
理
法
CUCUMBER WITH SEAWEED (*wakame*). Slice a peeled or unpeeled cucumber finely as for cucumber salad, place the slices in a bowl and sprinkle with table salt. After about 10-20 minutes the salt will have drawn the liquid out of the cucumber slices (an *osmotic* effect). Squeeze the slices gently with your hands and transfer to a dry bowl. In the meanwhile, allow the *wakame* to soak in warm water, rinse it thoroughly, and then cut it into pieces that are ca. 3 centimetres (1 inch) in size. Prepare a dressing from rice vinegar and soy sauce. Mix all ingredients together and adjust the seasonings with sugar and salt according to taste.

CUCUMBER WITH SEAWEED

This salad is the Japanese version of the type of cucumber salad distinguished by a sour-sweet dressing found in many cuisines. The salad consists of a mixture of soaked *wakame* seaweed fronds and slices of cucumber. It is easy to make and it keeps in the refrigerator for a couple of days.

As *wakame* can secrete a slimy substance consisting of *polysaccharides,* often used to set jellies, the rehydrated leaves have a tendency to stick to each other. To prevent this, wash them thoroughly so that it will be possible to mix them with the cucumber.

GHERKINS WITH *GARI*

Fresh gherkins are generally available for just a short period of time in the summer. Most people know them only as pickles, but if picked very small at the beginning of the season they are delicious eaten fresh with salt and a little pickled ginger, *gari*.

Gherkins with *gari* make a very special side dish or salad to go with sushi or *sashimi*.

調理法 **GHERKINS WITH *GARI*.** This recipe is based on a traditional Chinese recipe given to me by my good friend and physics colleague, Ling Miao. Clean fresh, small gherkins thoroughly in plenty of water, being sure to remove all sand and dirt. Cut off the ends which can be very bitter on some varieties, especially toward the end of the growing season. Make a deep cut lengthwise and one or two crosswise in each cucumber and then squash them lightly between two wooden cutting boards. They must not be crushed completely, just sufficiently for them to have cracked and broken surfaces. Sprinkle them with table salt and pour a little marinade from the *gari* over them as well, if desired. To finish, mix them with finely chopped pickled ginger.

Gherkins with *gari*.

Cucumber with seaweed (*wakame*).

SMALL DESSERTS WITH GREEN TEA

Desserts play only a minor role in the Japanese kitchen and none is really needed after a sushi meal. The few desserts which are found are not particularly sweet and are predominantly based on fruit and bean paste. I personally have a weakness for those made with green tea. The slightly bitter taste of the tea is a nice complement to ice cream and sweet fruit and its green colour shows up beautifully when combined with pale *tofu*.

Jellies with green tea

Many Japanese desserts and confections are based on a paste made from small, sweet *azuki* beans, both of the green and the red varieties. The red ones are the sweetest and are used to make the finest confections. One of these is a stiff jelly called *yōkan* which is eaten as a candy. Often *yōkan* is mixed with green powdered tea, *maccha*, and used as a filling in cakes and other confections that have a limited shelf life and must be consumed when very fresh.

As *azuki* bean paste is generally sold only in specialty shops, we have to make do with a number of substitutes. Here are two possibilities.

The easiest and most primitive way to substitute for the real thing is to buy small blocks of a firm jelly made from *kanten,* which is the gelling agent *agar* extracted from seaweed. *Maccha* has been mixed into these jellies, giving them a strong taste of green tea, but unfortunately they also contain a great deal of sugar. I often eat these as candy.

You can prepare a more elaborate jelly dessert from chestnut paste, which is available in cans. It can also easily be prepared from scratch from fresh chestnuts by soaking, cooking, and peeling them, and then puréeing them in a blender. The resulting chestnut purée is a good substitute for the bean paste.

Green tea jellies cut into thin slices and laid out attractively on a small platter make a simple dessert following a sushi meal.

Bean paste with green tea, *maccha*.

調 **Chestnut jelly** with green tea.
理 This jelly is based on a purée made
法 from chestnuts. Mix together 30
millilitres (2 tablespoons) of *maccha* dissolved in a little lukewarm water, 250 grams (1 cup) chestnut purée, 250 grams (1 cup) sugar, and ½ litre (2 cups) water and bring them to a boil while stirring constantly. Dissolve 15 millilitres (1 tablespoon) of gelatine powder in a little cold water and add the solution to the mixture in the pot. Pour the mixture through a sieve and put in a small rectangular metal pan.

The dessert is ready when the jelly has set completely. Dip the pan quickly in warm water to loosen the jelly, place a plate on top of the pan and unmold the jelly carefully by quickly flipping it over.

For a more colourful effect, make one portion of jelly with *maccha* and one without. Pour the first portion into the mold and add the second one after it has set. Serve the jelly uncut on a small platter and slice it at the table.

Red bean paste with whole *azuki* beans.

調
理
法 *Tofu* WITH GREEN TEA. This recipe calls for 1 litre (4 cups) of soy milk. Dissolve 30 millilitres (2 tablespoons) of gelatine powder in a little water. In a pot mix together the soy milk, 200 grams (¾ cup) of sugar, a little vanilla essence, ca. 50 grams (3 tablespoons) of *maccha* dissolved in a little lukewarm water, and the gelatine solution. Warm carefully, stirring constantly, and ensuring that nothing sticks to the bottom of the pot. Strain the mixture and pour it into a small rectangular metal pan.

The dessert is ready when it has set completely. Loosen the edges of the *tofu* by dipping the pan quickly in warm water and unmold it onto a platter. Serve the *tofu* in one piece and slice it at the table. As with the chestnut jelly, a combination of one portion of *tofu* with *maccha* and one without makes for an elegant study in contrasts.

TOFU WITH GREEN TEA

Tofu is produced from soy milk. It is a major task to make soy milk from scratch using beans and, as it is now widely available, it is much easier to buy it ready made. In a dessert, the bland taste of *tofu* can be counterbalanced by a combination of the sweetness from sugar and the bitter flavour from *maccha*.

Tofu with green tea.

HONEYDEW MELON WITH GREEN TEA

This dessert is an easy and quick way to round off a sushi meal. To please the taste buds, there is a finely balanced set of taste sensations – the sweetness and freshness of the melon cleanse the palate, but are offset by the slight bitterness of the *maccha*. And for the eyes, the three different shades of green – the honeydew melon, the tea, and the lemon balm – add visual appeal.

Melons (*meron*) are highly prized in Japan and those of excellent quality are also extremely expensive. Often given as presents, they are sold packaged in beautiful boxes, costing up to $200 U.S. for a single melon.

ICE CREAM WITH GREEN TEA

This is another handy dessert that requires little preparation. If you have a good quality vanilla ice cream in the freezer and a little green tea powder, *maccha*, this can be whipped up in a few minutes. The slightly bitter taste of the *maccha* is a nice counterpoint to the sweet taste of the ice cream. You can personalize this dessert and refine it further by making the ice cream yourself according to your favourite recipe and incorporating the *maccha* from the start.

調理法 ICE CREAM WITH GREEN TEA. Take the vanilla ice cream out of the freezer and let it sit in the refrigerator for about 10 minutes before it is to be used. Place the ice cream in a bowl and break it up into large lumps. Pour *maccha* powder on it while turning it carefully. This part is tricky: do not distribute the tea evenly, as you want a striped effect, but also make sure that you have no undissolved lumps of *maccha*. If you do not have *maccha*, you can pulverize a few leaves of green *sencha* tea in a mortar with a pestle. *Sencha* has a more bitter taste than *maccha*, so use a little less and experiment a bit to arrive at the right amount.

Should you wish to distribute the green tea evenly in the ice cream, you can first dissolve the powder in a little lukewarm water and mix it in when the ice cream has been allowed to go so soft that you can stir it in completely. In this case, you will need to refreeze the ice cream before serving.

Ice cream with green tea.

Melon with green tea.

調理法 MELON WITH GREEN TEA. A ripe, sweet honeydew melon which has green flesh works best. Cut it in half, remove the seeds, and peel. Cut the pieces into cubes and place them in a bowl. Dissolve 15 millilitres (1 tablespoon) of green tea powder, *maccha*, in a little cold water and pour it over the melon pieces, turning them carefully to coat them. Arrange the melon in a serving dish and decorate with a small piece of lemon balm.

AT THE TABLE
& AT THE BAR

A monk sips morning tea
it is quiet
the chrysanthemum is flowering
Matsuo Bashō (1644 - 1694)

How does one eat sushi?

When one eats sushi is it useful to think about its origins and the way it developed into *nigiri*-zushi and *maki*-zushi. While sushi was created in bite-sized pieces as an early form of fast food to be eaten with the fingers, it is still helpful to know a bit about how to handle chopsticks, especially for the side dishes that cannot be picked up with the fingers. And, of course, there are also a few subtle, unwritten rules.

The order of presentation in a sushi meal

In principle there is no set order of presentation for the different types of sushi and the side dishes. Hence, it is very simple; you do as you please. That being said, however, most Japanese would first eat the *sashimi*. And when it comes to the *nigiri*-zushi, they would probably select the tuna (*maguro*) first, especially if it is a fine, fat tuna (*toro*).

If sushi with *nori* is being served, it is a good idea to eat it first, or shortly after it has been prepared, because the *nori* absorbs moisture and quickly goes soft. *Nori* should be eaten while it is still crisp.

On a cold day, I would probably start with the soup, but it can be consumed at any point, also on an on-going basis throughout the meal, as long as it is still warm. Usually there is a lid on the soup bowl which will keep the contents hot for at least a little while. It goes without saying, however, that you also have to be careful to allow a very hot soup to cool a little. The soup is drunk from the bowl, and pieces of fish, vegetable, mushroom, and *tofu* are picked up with the chopsticks. In accordance with Japanese custom, you are welcome to make slurping noises when you are eating soup. It is even considered the polite thing to do in Japan, as it indicates that you appreciate the taste of the soup. In addition, it is a practical way to help cool off a soup that is very hot. But the rest of the meal is to be eaten soundlessly!

珍談 IT IS SAID THAT sushi connoisseurs always start with a piece of omelette (*tamago-yaki*) and judge the qualifications of the sushi chef according to how expertly it has been prepared. It is not easy to make *tamago-yaki* properly, whereas it is easy to slice a fillet of first-rate fish. If the *tamago* is good, the sushi bar passes muster.

珍談 IN JAPAN IT IS SAID THAT a stomach which is four-fifths full knows no doctor.

学術 **WHEN SUSHI FALLS APART.** If you follow all the rules and a piece of *nigiri*-zushi comes apart before you have placed it securely in your mouth, it really is the fault of the chef. Either the rice was not cooked right, the rice ball was not pressed together firmly enough, or the piece of fish was not pushed sufficiently into the top of the ball. Just try to do your best to salvage the remains without making a mess!

学術 **WASABI AND TEARS.** *Wasabi* is also called *namida* which means tears. Put a bit of freshly made *wasabi* paste on your tongue or hold it carefully near your nose or eyes and you will immediately know why.

HOW TO EAT SUSHI

Sushi can be eaten with the fingers, but most people, including the Japanese, use chopsticks. There are no hard and fast rules. For this reason nobody would look twice if you used your fingers to eat *nigiri*-zushi or *maki*-zushi. You can wipe the fingers off on the damp white washcloth (*oshibori*) which is normally given to you at a sushi bar before the meal starts.

Before beginning to eat, however, you need to pour a little soy sauce (*shōyu*) into the small flat dish or plate intended for that purpose. It is considered courteous to pour it for another if one is in company.

Wasabi is usually incorporated in the preparation of *nigiri*-zushi and *maki*-zushi, and I find that this produces the best results, provided that an appropriate amount is used. You can then fine-tune the taste by adding a little *wasabi* to the soy sauce and dipping the pieces in the sauce. When eating *sashimi*, which is prepared without *wasabi*, this is especially desirable.

A small amount of soy sauce is poured into a flat bottomed dish or onto a small plate. Pieces of sushi and *sashimi* are dipped in it very carefully and should never be splashed around in it.

Wasabi paste of a firm consistency formed into a small pyramid, ready for use.

Small bits of pickled ginger (*gari*) are eaten in between the different pieces of sushi and *sashimi* to cleanse the mouth and the palate.

Place a little *wasabi* paste at the edge of the small *shōyu* dish and, using the chopsticks, stir it gradually into the soy sauce until the mixture is the right strength for your taste. It can be a bit difficult to dissolve the *wasabi* in the soy sauce and the task requires patience. With a little dexterity, you can achieve the desired concentration of *wasabi* without stirring it into all the soy sauce in the dish, leaving some plain soy sauce on one side.

Dip *nigiri*-zushi in the soy sauce with the top side downwards. There are two reasons for doing it this way. In the first place, it ensures that the rice will absorb as little soy sauce as possible and there is less risk that the rice ball will fall apart. Secondly, the intention is that the soy sauce and the *wasabi* should adjust the taste of the fish or whatever else is used as topping. For the same reason, *nigiri*-zushi should be placed in the mouth with the rice upwards so that the fish lands on the tongue and gives the first taste impressions. You have to eat the whole piece in one go – biting it in two is not acceptable.

Rolled sushi (*maki*-zushi) and *gunkan*-zushi (battleship sushi) pieces are dipped into the soy sauce at the edge or the bottom of the piece. The sheet of *nori* is there to ensure that the rice stays together. Some *futomaki* rolls found outside of Japan can be so big and have so much filling that you really cannot eat a whole piece in one go and you must bite it in half. Hand rolled sushi (*temaki*-zushi) in the shape of a cone is eaten with the fingers; either dip the cone carefully in the soy sauce or cautiously pour a bit directly into it.

In between individual pieces of sushi, you can savour a bit of pickled ginger (*gari*). If you are a beginner proceed with caution as you might find the taste a little sharp, but eventually you will not be able to eat sushi without it. I sometimes enjoy eating *gari* almost like candy. It helps to cleanse the mouth and the palate to enable you to make the fine distinctions in the nuances of the tastes of the different fish and shellfish. You can likewise rinse the mouth with a sip of *miso* or *suimono*.

At this point we can no longer avoid a discussion of the chopsticks (*hashi*). *Sashimi* has to be eaten with them. For that matter, the side dishes, such as *tsukemono*, pickled ginger (*gari*), salads, and any warm dishes naturally cannot be eaten with the fingers either. There is no escaping the use of chopsticks.

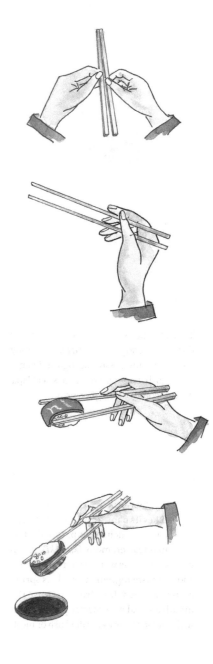

This is how you use Japanese chopsticks (*hashi*). Note that the piece of *nigiri*-zushi is dipped in the soy sauce top downward so that the rice ball is not dissolved.

Typical paper sleeves (*hashibukuro*) for disposable chopsticks (*waribashi*). They are often printed with the logo and name of the sushi bar or a simple, classical Japanese decorative design or print.

Disposable chopsticks (*waribashi*) that come in a paper sleeve (*hashibukuro*) are joined at the top and are separated by pulling them apart quickly. Use them as shown on the illustration, bearing in mind that this requires some practice. Most beginners have a tendency to hold on too tightly and squeeze the chopsticks. This makes it a great deal more difficult to control and manoeuvre them. It actually requires very little force to hold on securely to even a large piece of sushi. Try to hold the chopsticks as close to the top ends as you can manage with a light and relaxed grip. In contrast to the way a knife and fork are handled, which requires elbow movement, chopsticks are manipulated mainly with the help of the fingers and the wrist. A simple exercise that will help to acquire expertise in the use of chopsticks is to pick up individual grains of rice. It is much easier than it sounds.

When the chopsticks are not in use, you should place them on the holder (*hashi-oki*), which can be as simple as a flat stone or as elaborate as an artistic creation in ceramic or porcelain. If no chopstick holder is provided, it is a good idea to make one by folding the sleeve from the chopsticks into a little base. When you have finished eating, you lay the chopsticks horizontally across a bowl or plate and do not put them back in the paper sleeve.

学術 'CHOPSTICK WASH' (*hashiarai*) takes its name from an in-between course which forms part of the formal tea ceremony (*kaiseki*). It can consist of a little warm water to which is added a taste ingredient, such as spring onion, is added. It is drunk to rinse the mouth and palate, especially after a dish with a strong taste like grilled fish or meat.

Different types of chopsticks (*hashi*) and an assortment of holders (*hashi-oki*). The holder on the bottom right is made from a folded paper sleeve that covered the disposable chopsticks.

CHOPSTICKS AND THEIR 'TEN COMMANDMENTS'

There are more than fifty different types of Japanese chopsticks (*hashi* or *ohashi*). In contrast to Chinese chopsticks, they are pointed at the end which holds the food. Some are made from bamboo or soft, pale wood, others from elegant ebony, and still others from different metals. Many can be reused in the home, but the vast majority are disposable. Approximately 130 million disposable chopsticks are used every day in Japan, most made from imported wood. A substantial proportion of this wood is grown solely for this end-use, and the manufacture of chopsticks consumes a significant share of the world's wood resources. The most frequently used varieties are bamboo, willow, and cypress, while cedar is preferred for the high quality sets.

In classical times in Japan it was normal practice to have one's personal chopsticks in a little holder which could be brought along on journeys. In the same way as rice, chopsticks were a symbol of prosperity in the olden days.

If you want to observe Japanese customs, you absolutely cannot do as you please with your chopsticks. The rules which follow fall into two categories: some are simple and practical, while others are more ritualistic and strongly culturally influenced.

YOU MUST NOT

1. Lick or suck on the chopsticks.

2. Stand the chopsticks in a bowl filled with rice. This is a ritual that belongs on a home altar.

3. Move bowls or dishes around on the table with the chopsticks.

4. Rub the chopsticks repeatedly against each other (this signifies that they are cheap and of poor quality).

5. Take from a common plate without using the special chopsticks for that use. If there are none, turn your own chopsticks around and use the other end for serving food. This rule is *Shinto* inspired. According to a *Shinto* way of thinking, an evil spirit can move from one person to another via the food.

6. Hold the chopsticks in mid-air in front of you or over the food while you are deciding what to pick up.

7. Drop food or soy sauce from the chopsticks.

8. Let the chopsticks rest on bowls during the meal. They are placed together on the chopstick holder (*hashi-oki*) or on a little base which you can fold together from the paper sleeve that covered them.

9. Spear the food with a chopstick or eat with one only.

10. Transfer a piece of food from one set of chopsticks to another. This is because, in Japan, after a person is cremated the bone remnants of the deceased are transferred from one set of chopsticks to another before the ashes are placed in the urn.

Masu – a measuring cup made from wood, used for drinking *sake*. If you have never tried this before, you have to reflect a bit on whether you want to drink from the side or from the corner.

学術 **SAKE IS WARMED** in a water bath. The *sake* carafe is filled to about three-quarters of its capacity and placed upright in a pot which is half filled with water near the boiling point. The correct serving temperature for *sake* is 35-40°C (95-105°F).

学術 **A TEA CONTRETEMPS.** In a well-known sushi restaurant in Denmark, I was once informed by a very over-bearing server that warm green tea was definitely not something that one could drink with sushi. My American guests were flabbergasted and, needless to say, I have never been back to that restaurant.

WHAT SHOULD ONE DRINK?

The answer to this question is simple but, nevertheless, complicated. Many gastronomes have gone to great lengths to discover the 'right' wine to serve with Japanese food, especially sushi. My solution is straightforward: forget about wine and drink green tea, beer, or *sake*. Many Japanese people feel that *sake* is the ideal accompaniment for fish. But others will say that *sake* should be drunk before the sushi meal because it is made from rice and drinking rice wine with a meal that has a significant rice content is completely superfluous.

If only one beverage is to be served, I would choose green tea. Its mild, rounded, slightly bitter taste goes well with vinegared rice and the subtle flavours of raw fish. Green tea is also an effective thirst quencher – both the thirst you might feel at the beginning of the meal and the thirst inevitably brought on by the salt in the rice and the soy sauce. In addition, green tea, being very slightly astringent, effectively cleanses the mouth from the taste of oily fish. The tea does not need to be really hot and, in fact, green tea tastes best when it is not overly warm. It is usually drunk from a *raku* ceramic mug or a fine porcelain cup, both of which have no handles.

In addition, of course, there is beer. Japanese beer is first-rate, mild, and not as bitter as most European-style lagers. I enjoy unfiltered wheat beer with sushi, as it is mellow with sweet undertones and also quenches one's thirst.

All this being said, a formal, traditional sushi meal is always accompanied by *sake*. While I personally think that it should be drunk warm, opinion is divided on this subject. *Sake* is served in small bowls, *sakazuki*, or little cedar boxes, *masu*. The latter are derived from a dry goods measurement, 180 millilitres, that originally corresponded to the amount of rice needed to make an individual portion. Be sure to empty the *sake* bowl before you eat any sushi or *sashimi* and not afterwards.

Politeness dictates that one should pour *sake* or tea for each other. Filling one's own cup is interpreted as greed.

On occasion, Japanese people decide to toast each other in the course of the meal – a ritual very similar to the one widely used in the Western world. In Japanese, the word used to propose a toast is *kampai*.

CHA – JAPANESE TEA

Japanese green tea, *cha*, is much more than a beverage
which is a suitable accompaniment for cooked Japanese
food or sushi. Green tea is deeply rooted in the Japanese
culture and way of thinking and it is the central point
around which the renowned Japanese tea ceremony revolves. There are many different types of green tea, each
with its own character and taste, selected to suit the occasion. Green tea tastes perfect with sushi.

TEA IN JAPAN

Drinking tea in Japan is not just a question of special types of tea, where they are grown, and which water is used to brew the tea. Both the tea and how it is served are shrouded in sophisticated rituals and are deeply symbolic. You will probably see only a little, or possibly none, of this at a typical sushi bar in the West. On the other hand, you are likely to notice the care with which the tea is served and that the tea cups are often attractive in their own right.

But Japanese green tea is so much more. It is a narrative about how to perfect simplicity, cleanliness, beauty, harmony, and respect for things and people, . It is the embodiment of *wabi sabi*. It is also the history of famous tea masters, tea schools, tea houses, and meals served with tea (*kaiseki*). It encompasses the rituals associated with the preparation and serving of tea (*chanoyu*), as well as the special relationship between the guest and the host at a formal tea ceremony.

In comparison to the simplicity of tea and the ease with which it can be prepared – this really amounts to nothing more than pouring boiling water on some leaves from a bush – the cultural history bound up with tea is remarkably rich.

Tea in the form of powdered tea leaves (*maccha*) is thought to have arrived in Japan in the 9th Century, when it was brought from China by monks and traders. Without a doubt, the stimulating effects of the tea helped the Buddhist monks during their long prayers and religious observances. But it was only from the conclusion of the 12th Century that tea started to be cultivated in Japan and to be prepared from tea leaves (*sencha*). It was grown at the Buddhist temples and the Japanese monks were the first to elaborate and refine the *Zen* inspired methods and rituals linked to the preparation, serving, and drinking of tea. At this time the dried tea leaves were ground to a fine powder (*maccha*) that was combined with boiling water and whisked into a thick tea using a small bamboo brush. This was the foundation of the tea ceremony.

In principle, there are two types of tea: green tea and black tea, both of which are produced from the leaves of the tea bush, *Camellia sinensis*, or its varietals. But in Japan and with Japanese food, it is overwhelmingly green tea that is consumed.

珍
談
㐧

IT IS SAID THAT drinking tea was invented in China more than 5000 years ago by the legendary, scientifically curious emperor Shen Nung who had decreed that all drinking water must be boiled for reasons of hygiene. On one of his journeys to the far reaches of his domain, some dry leaves fell, quite by accident, into the water which his servants were boiling for him. The emperor, who became fascinated by what was being extracted from the leaves, tasted the water and found it refreshing. Tea had been invented.

Tea bushes.

Japanese confection with soybean paste to which powdered green tea (*maccha*) has been added.

学
術 Is green tea good for you? The answer is probably yes. In the first place, there is a significant quantity of *antioxidants* in green tea. Furthermore, recent research has shown that the principal type of *polyphenol* (epigallocatechin gallate) in green tea can counteract cancer, especially stomach and throat cancers. Apparently, the mechanism involved is that the polyphenol suppresses the growth of blood vessels in the tumor.

While black tea is subjected to extensive processing before it ends up in the teapot, green tea is produced from green tea leaves and handling of the leaves is kept to the minimum required to make the leaves last until they are used.

Green tea accompanies every Japanese meal, not least of all sushi. Tastewise, green tea goes well with fish and rice, and it is a good way to quench the thirst brought about by eating the salty soy sauce. A Japanese meal always concludes with green tea. Asking for a little more green tea at a sushi bar when you do not feel like eating any more is an excellent way to signal to the chef and the server that the meal is finished.

In addition, green tea is used to flavour ice creams, candies, and confections made from soybean paste.

Green tea and black tea

Green tea makes up one quarter of the world's production of tea. While it is supposed to consist of the youngest and freshest leaves, somewhat older and larger leaves are mixed into the cheaper versions. Very soon after the leaves are picked, they are warmed to 95°C using steam (or, as in China, toasted on a warm pan) to *denature* the *enzymes* that would otherwise begin to *oxidize* the leaves' essential taste substances, the *phenols*. Afterwards the leaves are rolled at a temperature of about 75°C and then dried.

Phenols, which make up about 25-35% of the dry weight of the fresh, green leaves, are the source of the tea's characteristic slightly bitter taste with notes of grass, hay, flowers, and seaweed. In addition, tea leaves yield an amino acid, *theanine*, which is converted to the slightly sweet and spicy *glutamic acid*.

Substances called *flavonols* and *flavones* are the source of the green or yellowish colour of the tea and also help to amplify the bitter and harsh taste imparted by the *phenols*. In addition to these important ingredients, several hundred different olfactory substances have been discovered in green tea.

Green tea is best preserved in an airtight package and, unfortunately, it loses its aroma if it is stored for too long a period of time. I often resort to wrapping the tea leaves up tightly and freezing them to preserve their flavour.

In the West, black tea is much more popular. In contrast to green tea, black tea has undergone extensive processing, consisting of withering, rolling, *enzymatic oxidation*, firing, and drying of the leaves. The enzymes found in the tea, *polyphenol oxidase*, are active in the oxidation process. It results in a whole series of aromatics and pigments that are complex chemical compounds derived from the *phenols* originally present in the green tea leaves.

It is often said that black tea undergoes a *fermentation process*. This is not correct as no microorganisms are involved, only the tea's own enzymes. The chemical compounds created in black tea have dark colours, are less bitter, and have a more rounded taste than the phenols found in green tea.

珍談 It is said that the second steeping of green tea is the best. This is also my experience. As green tea contains only small amounts of *astringent phenols* (*tannin* or *tannic acid*), using the tea leaves a second time does not result in a brew that is significantly more bitter than the first and often in one with a richer taste.

Water for tea

Volumes have been written about the proper way to make green tea – in what type of water should the leaves be steeped, at what temperature, and for how long?

Here are some dry facts. First, the question of the hardness and acidity of the water. Very hard water which has a high mineral content of calcium, magnesium, and iron is less effective for making tea because a foam of *phenols* and *calcium sulphate* is deposited on the surface of the tea leaves, inhibiting the steeping process. Very soft water, on the other hand, is too effective and the resulting tea has a salty taste. Hence, the golden mean is also the right way to make good green tea. As green tea is of itself a bit acidic, it is best if the water used to make it is neutral or a little alkaline.

Next comes the question of the right temperature. It is well known that one makes the best coffee and, for most types of black tea, also the best black tea if the water is near the boiling point and the temperature is maintained during the entire steeping process.

Traditional Japanese water kettle.

Depending on the variety of green tea used, the best result will usually be obtained by steeping the tea at somewhat lower temperatures. Typically these are from 50°C (120°F) to about 90°C (195°F) for a period of from half a minute to a few minutes. When in doubt, it is better to use too cool than too hot water. Let the tea stand for a minute after it has steeped to allow the flavours to develop fully.

Under no circumstances should the water be boiling when you make green tea from the better varieties. Nor should the water have boiled before it is used. It should be fresh – the oxygen should not have been driven off and the minerals in it should not have precipitated out.

This is related to a very poetic story about how one determines when the water has reached precisely the right temperature. It is a poetic story with a scientific explanation to back it up.

In traditional Japanese literature about the noble art of making tea, the boiling of water is divided into three stages. The first sign that it is boiling is the formation of bubbles on the bottom of the kettle, 'fish eyes', at 70-80°C (160-175°F). The 'fish eyes' are accompanied by a faint hissing sound. At the next stage, the bubbles form strings, 'pearl necklaces', from the bottom to the top of the kettle. This happens at 80-90°C (175-195°F) and one hears bubbling and humming in the water. The final and most intense stage is evident when large steam bubbles break the surface. Now there is *turbulence* and a rumbling sound.

学術 **ASTRINGENCY IN TEA.** *Astringency* is a harsh, biting physical feeling in the mouth which is not actually a true taste sensation. We best know astringency from black tea that has steeped too long or from red wine which contains too much *tannic acid. Tannins* (also called *tannic acid*) are *phenols.* The tannins bind the *proteins* in saliva causing a dry, rubbing sensation and this is how we experience astringency. The sensation can be either pleasant or unpleasant, depending on the context.

学術 **MORE CAFFEINE IN TEA** than in coffee? Tea leaves contain 2-3% caffeine, while coffee beans have only 1-2%. But one uses more coffee than tea for a cup, so there is normally more caffeine in the former than in the latter. Caffeine is an *alkaloid* which stimulates the central nervous system and also affects the heart beat and blood circulation.

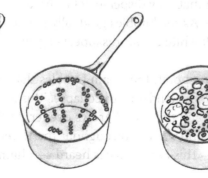

Fish eyes   Pearl necklaces   Turbulence

WHY DOES THE KETTLE 'TALK' WHEN WE BOIL WATER?

We have all wondered about the sounds emanating from the kettle when we boil water. First there is a hissing sound for some time before the water boils, then a slightly deeper and fainter sound, and finally, when the water boils, a bubbling and rumbling sound. The more curious among us have probably also noticed that the water becomes cloudy and milky before it reaches the boiling point.

We are not the only ones who have taken an interest in this aspect of kitchen physics. The Scottish physicist Joseph Black (1728-1799) did as well, so he went further and found a scientific explanation for the phenomenon.

When one heats water, the first thing that happens is that air, which is dissolved in the water, collects in small bubbles on the bottom of the kettle. When these air bubbles tear themselves loose from the bottom, it gives rise to a faint, deep sound (100 Hz) that is barely audible. Closer to the boiling point, small steam bubbles are formed at the bottom of the kettle. This is the *fish eyes* stage and the beginning of the *pearl necklace* stage. The steam in the bubbles is hotter than the surrounding water and when the bubbles start to rise up through the water they are cooled, causing the steam pressure inside the bubbles to decrease.

When the steam pressure inside the bubble is no longer sufficient to withstand the surface tension of the bubble, it implodes, collapsing just like an explosion in reverse. The implosion results in a sudden displacement of the water, which then falls into the bubble, and this displacement is like a sound wave. Mathematical calculations show that the frequency of the sound wave is about 1000 Hz, equivalent to what we register as a hissing sound. When the bubbles collapse, they do not disappear completely. Some much smaller bubbles are left, which among other things contain air. These bubbles are actually so small that they disperse light; therefore, the water is no longer clear but looks milk white. As the bubbles gradually rise to the surface and the water becomes hotter, the milk white colour disappears again.

When the water is very near the boiling point, the steam bubbles that are formed on the bottom become larger and have less tendency to collapse. This is observable from the fact that the kettle 'talks' less and the sound grows deeper. At the boiling point, bubbles are formed both at the bottom and throughout the water, and they break through the surface of the water. This *turbulence* is heard as a humming and roaring sound.

Gyokuro

Sencha

Bancha

Hojicha

Genmaicha

Maccha

JAPANESE GREEN TEA

Gyokuro is considered to be the finest Japanese green tea, and it is also the most expensive. Only the leaves from young tea bushes are used and they are shielded from the sun during the last few weeks of their cultivation. This causes slower growth, preserving their freshness and allowing fewer bitter taste substances to develop. In this way, the taste of *gyokuro* retains a certain sweetness, preventing bitterness from dominating. Only the very best outer leaves are utilized – they are shiny and have a beautiful, deep green colour. The tea should be steeped in water that is at 50-60°C (120-140°F) for two to three minutes.

Sencha is a good quality tea that one should not be ashamed to serve to guests. It should be steeped at ca. 80°C (175°F) for two minutes. It is possible to brew more than one batch from the same leaves. The first steeping is the most aromatic, the second is sweeter, and the third is overwhelmingly bitter. I prefer the second steeping, which has a suitable, complex combination of aroma, sweetness, and bitterness. *Sencha* tastes a little the way fresh hay smells. It is the perfect tea to accompany sushi, neither too delicate nor too rough.

Bancha is the everyday tea, usually served at breakfast. The leaves are of a coarser quality than those for *sencha* and some mixtures include twigs. *Bancha* is steeped in boiling water for less than a minute.

Hojicha is *bancha* that has been fired at 180°C. The firing concentrates taste and aroma and gives a less bitter brown tea, which contains less caffeine than green tea and has a slightly smoky taste. *Hojicha* is steeped for under a minute in boiling water.

Genmaicha is a mixture of *bancha* and toasted rice grains that have been puffed like popcorn. This tea, which has a slightly nutty taste, should also be steeped in boiling water for under a minute.

Maccha refers to a finely powdered green tea, often sold in tins. This is the tea that is used for the formal Japanese tea ceremony and with *cha-kaiseki*. The tea powder is dissolved in water that is at a temperature of about 80°C (175°F). *Maccha* powder is also well suited for making desserts, for example, with fruit or ice cream.

Sencha

Genmaicha

Hojicha

Maccha

学
術

Thin green tea and thick green tea refer, respectively, to a drink made by steeping tea leaves in warm water and to a drink made by whisking green tea powder together with warm water (*maccha*).

All types of green tea are drunk without milk or sugar.

Maccha etiquette

Maccha is drunk in accordance with a unique, formal set of rules. One holds the drinking bowl in the palm of the left hand and places the palm of the right hand against it with a little flourish. If one is the host, one must turn the bowl clockwise so that the front of the bowl faces the guest, as it is quite possible that this side has an especially beautiful decoration. The guest receives the bowl and then must turn it clockwise by ninety degrees twice in a row; this leaves it with the front facing the host. The guest then drinks from the back of the bowl.

Etiquette prescribes that the tea must be drunk in precisely three and a half gulps. No more and no less.

調
理
法

Whisked powdered green tea – *maccha*. For one cup of *maccha* place 5 to 10 millilitres (1-2 teaspoons) of *maccha* powder in a bowl that has been heated in warm water. Add 100-150 millilitres (about ⅔ cup) water at a temperature of about 80°C (175°F). Holding the bowl in one hand, mix the tea together with the other hand using the bamboo whisk (*chasen*) so that the tea starts to froth. The whisk must be soaked in water before use so that the tea powder does not clump together on the fine strands of the bamboo. Be sure to whisk thoroughly around the edge of the bowl. The tea is then poured into a cup or drinking bowl and drunk at once while there is still a little foam on its surface.

Equipment for the tea ceremony: drinking bowl, bamboo spoon for measuring tea powder (*maccha*), bamboo whisk (*chasen*) for mixing the tea together, and cloth (*fukin*) for wiping the edge of the bowl when more than one person is to drink from it.

Ichigo ichie – each moment, only once

On the tea ceremony

The tea ceremony, *chanoyu*, is an example of Japanese culture in which history, ritual, socialization, and mystical religious philosophy come together on a higher plane. It is virtually impossible for an outsider to be able to describe properly what the tea ceremony really consists of, even less to gain an insight into the way of thinking that underpins it, or to learn how to perform it correctly. Seemingly, it is no easy task for the Japanese to do this either and there are tea schools with longstanding traditions where tea masters give courses of instruction in the tea ceremony that last for years, if not for a lifetime. Young Japanese women still follow mini-courses in the tea ceremony as part of their wedding preparations. Here they learn the choreography of the ceremony, how to boil the water, how to whisk the tea, and how to serve it properly. It looks very simple on the surface, but only to the uninitiated who do not discern the deeper levels of meaning.

It is said that there are so many layers in Japanese culture and way of thinking that if one has lived in Japan for a month, one can write a book about it; if one has lived there for a year, one can write an article about it; and if one has lived there a whole lifetime, one can say only a single sentence. If this saying is true, my description of the Japanese tea ceremony should be very, very long. In her beautiful and empathetic book, *Untangling My Chopsticks*, Victoria

Abbott Riccardi describes how she, as a young Western woman, a foreigner, and an outsider, arrived in Kyoto in order to learn the tea ceremony. I have viewed the Japanese tea ceremony through Riccardi's carefully polished lens.

The tea ceremony builds on elements of the mystical *Zen* philosophy and its ideals of *wabi* and *sabi*, purity and simplicity. It was the *Zen* monk Murata Jukoh (1423-1502) who refashioned the more worldly tea parties of earlier times into a well-defined rite with a religious and philosophical orientation. In this way, the tea was transformed into something more than a beverage; it became an entity enclosed by rules and rituals. The rituals adhere strictly to the *Zen* admonition that you may have only one chance in a lifetime to do something, so you must govern yourself accordingly, be well prepared, and do it correctly. *Ichigo ichie* – each moment, only once. The tea ceremony, *chanoyu*, as it is practiced today was elaborated by Sen-no Rikyu (1522-1591), the most famous tea master of all time. Rikyu's three grandsons also became great tea masters and each founded a tea school in Kyoto.

Chanoyu means 'the warm water of the tea'. The idea is that the tea should be prepared in a manner and drunk in circumstances that foster meditation on life. The process itself is the end. It is based on the Buddhist concept of reaching the center of the mandala by detours and through concentration. The journey gives rise to the harmony and insight which are the ultimate goals, rather than the center itself. It is about harmony (*wa*), respect (*kei*), cleanliness (*sei*), and purity (*jaku*). Rikyu established seven rules for achieving this through the tea ceremony. 1. Make the flower arrangement (*ikebana*). 2. Burn wood charcoal to heat the water for tea. 3. Serve precisely the right amount of tea. 4. Cool the surroundings in summer and heat them in winter. 5. Prepare everything well ahead of time. 6. Be prepared for rain. 7. Guests and host must demonstrate mutual respect.

In accordance with these rules each detail and every scene is taken into account and choreographed to suit the circumstances – from the tea room and its appointments with the flower arrangement and the equipment used to boil the water to the actual preparation and serving of the tea.

Originally *chanoyu* was accompanied by either no food at all or else a very minimal meal – a bowl of rice, a little soup, and a few other vegetarian dishes. This mirrored the simple life in the temples of Kyoto. Over time, a more comprehensive, formal meal, *cha-kaiseki*, gradually evolved and was served by the monks for aristocratic and more demanding guests in the temple. In this way, the ritual acquired the aura of a truly festive occasion with a group of people eating together. *Kai* can mean 'group' and *seki* 'gathering place'. The tea ceremony, *chaji*, which follows the meal is, however, still the main event. *Kaiseki* can also mean something along the lines of 'warm stone'. This connects *kaiseki* with the word *yakuseki* which means 'medicine stone' and refers to the *Zen* monks' practice of holding a warmed up stone against their stomachs during the long and exhausting religious temple rituals in order to keep warm and to suppress hunger pangs. From a *Zen* point of view, the 'medicine stone' is regarded as the means of healing the disease of hunger.

Cha-kaiseki consists of a large number of dishes, principally rice, soup, grilled and cooked fish, and salt cured and pickled vegetables, *tsukemono*. The meal is elaborated in great detail and is heavily ritualized, varying in accordance with the season of the year and the specific circumstances. Only a little of each dish is served – not too much and not too little. This is *Zen*, for you are supposed to leave the meal slightly hungry in order to remember the food and to have a desire to come another time.

The tea ceremony itself, which follows the meal, takes place in another room or in a specially appointed, primitive tea hut. The path between the two must be aesthetically pleasing, preferably with water flowing past it.

Walking along it will put the soul in a state of readiness for the tea. The tea is prepared and served by the host. The wood charcoal used to heat the water is rearranged several times and the flame itself and even the way it is burning are regarded with veneration by the guests. First the 'thick tea' is served. The tea powder, *maccha*, is whisked into the warm water and served for the guests in a bowl from which they all drink. Next the 'thin tea' is prepared, brewed using tea leaves. This tea is served individually in lovely cups or bowls, the beauty, age, and form of which are duly admired.

There is little talk during the tea ceremony, as the guests in their kimonos sit on the floor for several hours, respectful of each other and the host, and with *Zen* in their souls. The *Zen* teaching about *mu* – signifying 'emptiness' or 'nothingness' – is the very heart of the tea ceremony.

The tea ceremony can appear complex, overly refined, and tedious. But in one of his poems about it, the great tea master Sen-no Rikyu reminds us that the tea ceremony consists of nothing more than boiling water, making tea, and drinking it.

Epilogue

You have ordered the tea and your meal has come to an end. Hopefully, the sushi was to your satisfaction and you enjoyed the food. Better yet, you might still feel a tiny twinge of hunger so that you will soon come back for more.

Thank you for joining me on a tour of the history, craft, art, and science of sushi. Thank you for allowing me to share with you my passion for sushi and some scientific insights into why there are so many health benefits from eating it.

I hope that my story has stimulated, challenged, and encouraged you to explore on your own, 'steal from the master', and engage yourself in searching for new sushi experiences and experimenting with it in your own kitchen. I did not tell you everything; there is much more out there for you to discover and invent.

The sushi food culture may be old and venerable, steeped in its own traditions and etiquette. It may be composed of ingredients that are rich in subtle flavours and aromas and others that may seem strange to some. Special equipment and serving pieces may be used for its preparation and presentation in esthetically pleasing arrangements.

But let me remind you that, when all is said and done, the art of sushi consists of nothing more than slicing raw fish, boiling rice, and eating the two together.

The technical details

Plum-blossoms everywhere
I should go south
I should go north
Yosa Buson (1716-1783)

しゃりはすしになくてはならび、水加減、切り方（混ぜ方）、ってもすしをおいしく握るためいる。このしゃりのでき次第でりと、しかも形をくずさずに握決まってくる。できの悪いしゃづらいのはもちろんのこと。おざわでは、米は茨城産のひを、水はアルカリイオン水を使飯に合わせるすし酢の加減も具合が適切でないと、ネタがいよさが生きてこないのである。つやも大切な一要素。そもそも酢という色のついた酢を使って在はしゃりの白銀色を出すため

GLOSSARY OF JAPANESE WORDS

agari – sushi bar slang for tea.

aka-jiso – red *shiso*.

akami – red and dark items (*tane*) used for sushi and *sashimi*, e.g., tuna; (*ako*, red).

aka-miso – red *miso* made from rice.

amaebi – sweet shrimp; (*amai*, sweet).

amakuchi – sweet *sake*.

anago – sea eel.

aoaka-jiso (*hojiso*) – green-red *shiso* with leaves which are red or dark purple on the surface and green on the underside.

ao-jiso – green *shiso*; (*ao*, green).

ao-nori – flakes of green seaweed.

awase-zu – mixture of rice vinegar, salt, and sugar which is added to cooked sushi rice.

azuki (*aduki*) – small green or red beans used as a paste in Japanese cakes, confections, and desserts.

bancha – ordinary green tea for everyday use made from coarser, larger tea leaves; may contain some twigs.

basashi – *sashimi* made from raw horse (*uma*).

battera – pressed sushi with mackerel, a type of *oshi*-zushi which is a specialty of Osaka.

bentō – meal arranged in a box divided into sections, usually including rice, *tsukemono*, and assorted other small dishes.

buri – see *hamachi*.

cha – Japanese green tea (*Camilla sinensis* or *Thea sinensis*) also called *o-cha*, which signifies that the tea is not enzymatically fermented (as is black tea, *kōcha*). The different types of *cha* are *gyokuro*, *sencha*, *maccha*, *bancha*, and *hōjicha*.

chaji – combination of tea ceremony (*chanoyu*) and the formal meal (*cha-kaiseki*).

cha-kaiseki – the formal meal served before a tea ceremony.

chanoyu – 'the tea's warm water', the formal way of preparing and drinking whisked green tea (*maccha*) at a tea ceremony.

chasen – bamboo whisk for mixing green tea powder (*maccha*).

chirashimori – see *moritsuke*.

chirashi-zushi – scattered sushi (also called *bara*-zushi), a particularly colourful type of sushi in which fish, shellfish, and green items are placed in a bowl on top of a layer of sushi rice (Tokyo style), on which finely cut *nori* and a little *tobiko* roe are sometimes sprinkled. *Gomoku*-zushi ('five ingredients sushi') is another type of *chirashi*-zushi, characteristic of the Osaka area, in which cooked green vegetables and the other ingredients are mixed together with the rice.

daikon – large, white radish (Chinese radish).

daizu – green soybean (*Glycine maximus*) used for making *tofu*, *miso*, and *shōyu* among other products.

dashi – fish stock made from bonito fish flakes (*katsuobushi*) and *konbu*. First *dashi* (*ichiban dashi*) and second *dashi* (*niban dashi*) refer to the first and second extract of *katsuobushi*.

deba-bōchō – Japanese knife for cutting up fish and shellfish.

denbu – see *oboro*.

donko – much sought-after *shiitake* mushroom with a small, dark cap.

ebi – shrimp; but the term encompasses a long list of similar crustaceans of varying sizes.

edamame – young, green soybeans (*daizu*) containing two to three beans in each pod.

Edo – former name of the city of Tokyo; also associated with the so-called Edo period which started in 1603 when the shogunate moved from Kyoto to Edo.

edomae-zushi – *nigiri*-zushi. Originally sushi made from fish and shellfish from the bay by Edo, the earlier name for Tokyo; now used to denote sushi of high quality.

enokitake – winter mushroom (*Flammulina velutipes*) with long, thin white stalks and a small cap; grows in a cluster.

fu – wheat gluten, also known by its Chinese designation *seitan*, in either raw form (*nama fu*) or roasted or dried (*yaki fu*).

fugu – pufferfish or blowfish of the *Tetraodontiformes* family. Liver and ovaries of the fish contain the potent nerve toxin *tetrodotoxin*.

fukin – cloth for wiping or drying.

funa-gata – see *nigiri*-zushi.

funamori – see *gunkan-maki*.

funa-zushi – sushi made from the carp *Carassius auratus*, a wild goldfish which lives in Lake Biwa close to Kyoto.

furikake – condiment often sprinkled on warm rice and other dishes; consists of a mixture of salt, dried bits of seaweed, and fish flakes, as well as toasted black or white sesame seeds.

futomaki – thick *maki*-zushi rolls made using a whole sheet of *nori*.

gari – sushi bar slang for pickled ginger (*tsuke-mono*) when it is associated with sushi, usually sliced very thin. The Japanese word for ginger is *shōga*.

genmaicha – tea mixture consisting of ordinary green tea (*bancha*) and roasted rice kernels.

geta – classical Japanese wooden shoe. The term is also used for the simple wooden block with feet used as a plate for sushi and *sashimi*.

goma – sesame seeds (*Sesamum indicum*); can be white sesame seeds (*shiro goma*) or black sesame seeds (*kuro goma*).

gomai oroshi – five part filleting of fish into four fillets and the remaining skeleton; used for flatfish such as turbot and large rounded fish like tuna.

gomoku-zushi – see *chirashi*-zushi.

gu – filling placed in *maki* rolls or mixed into *chirashi*-zushi.

gunkan-maki – sushi made by enclosing ingredients which might otherwise fall apart in a piece of *nori*; also known as battleship sushi (*kakomi*-zushi or *funamori*).

gyokuro – green tea of the very best quality.

hagotae – tooth resistance.

haiku – minimalist Japanese style of poetry governed by a set of complicated rules (*hai*, entertainment, and *ku*, fragment). Typically the poem consists of three lines with 5, 7, and 5 syllables respectively.

hako-gata – see *nigiri*-zushi.

hako-zushi – slices of raw fish placed between layers of cooked vinegared rice pressed together in a small wooden box for about 24 hours and then eaten immediately thereafter; forerunner of the more modern pressed Osaka sushi, *oshi*-zushi, which is usually made with mackerel.

hamachi (*inada, buri*) – yellowtail, a fine textured fish well suited for sushi and *sashimi*.

hangiri – wooden bowl for cooling newly cooked sushi rice.

hashi – chopsticks.

hashiarai – 'chopstick wash'; refers to an in-between course at the formal tea ceremony (*kaiseki*) where a little warm water, to which some flavouring has been added, is served and drunk to cleanse the mouth and the palate following a dish with a strong taste such as grilled fish or meat.

hashibukuro – paper sleeve enclosing disposable chopsticks (*waribashi*).

hashi-oki – small holder on which the chopsticks (*hashi*) are placed.

haya-zushi – sushi based on cooked rice mixed with rice vinegar and then kept under pressure with a stone weight and fermented in a wooden box over a short period (24 hours).

hijiki – brown seaweed (*Sargassum fusiforme*).

hikari-mono – shiny things (*tane*) which are placed on sushi, such as mackerel and herring which have their silvery skin left on.

hirame – flatfish which have the eyes on the left side, e.g., brill and turbot. Flatfish with eyes on the right side are called *karei*. The classification is ambiguous.

hiramori – see *moritsuke*.

hōchō – Japanese kitchen knife, available in various versions each with its specific use, for example, *yanagiba-bōchō* for slicing *sashimi*, *deba-bōchō* for cutting fish and shellfish, and *usuba-bōchō* for cutting vegetables.

hojicha – roasted green tea (*bancha*).

hokanomono – things (*tane*) placed on sushi that are not included in the classical categories of *akami*, *shiromi*, *hikari-mono*, and *nimono-dane*.

hone nuki – tweezers.

hon-maguro – bluefin tuna.

hoshi-nori – dried nori.

hosomaki – thin *maki*-zushi rolls made with a half sheet of *nori*.

hotategai – scallop (*Patinopecten yessoensis*).

ichiban dashi – the first *dashi*.

ika – squid.

ikebana – flower arrangement; originally a ritual way of arranging flowers (*bana*, flower) in Japanese temples where it was practised as a meditative art form through which one could cleanse the soul and find harmony and balance. *Ikebana* also encompasses stems and leaves, as well as the container in which they are placed. *Ikebana* is characterized by a linear, simple appearance and asymmetry.

ikijime – fish that are eaten immediately after they die and before *rigor mortis* sets in, so that their flesh has a firm texture which has not become tender as a result of natural decomposition. Typically white fish are used and they are killed in iced saltwater to limit their struggling. The fish are kept in tanks at the sushi bar, killed, and cut up on the spot.

ikizukuri – *sashimi* sliced from a whole, freshly killed fish and replaced decoratively on the fish skeleton before being served.

ikura – salted salmon roe.

inada – see *hamachi*.

itamae – he who stands 'in front of the cutting board', Japanese head chef. A sushi chef is referred to as a *sushiya*.

kaiseki – see *cha-kaiseki*.

kaiten-zushi – sushi served on a conveyor belt in a specially constructed sushi bar.

kaki – oysters (*Crassostrea gigas*).

kakomi-zushi – see *gunkan-maki*.

kampai – 'Cheers!' when proposing a toast.

kani – crab.

kanji – Chinese characters used in Japanese written language.

kanpachi – great amberjack or rudderfish, closely related to *hamachi*.

kanten – agar, a polymer of galactose, a polysaccharide derived from seaweed.

kappa – cucumber when related to sushi.

kappa-maki – *maki*-zushi with cucumber.

karakuchi – dry *sake*.

karei – flatfish which has the eyes on the right side, e.g., lemon sole, Greenland halibut, and halibut. Flatfish with the eyes on the left side are called *hirame*.

kasanemori – see *moritsuke*.

katsuo – skipjack tuna (*Katsuwonus pelamis*), also known as bonito; member of the mackerel family.

katsuobushi – cooked, salted, dried, smoked, and fermented *katsuo*, which is shaved into paper thin flakes; used to make such things as fish stock, *dashi*.

kazunoko – roe from herring (*nishin*).

kensho – Japanese *Zen*-related term for enlightment experiences.

kihada – yellowfin tuna.

kodomo-zushi – children's sushi or family sushi, typically *maki*-zushi with cheerful, colourful patterns in the cross-sections of the rolls.

kohada – gizzard shad (*Clupanodon punctatus*), also known as *konoshiro*.

kōika – cuttlefish (*Sepia esculenta*).

kōji – fermentation medium made from rice, barley, or soybean paste inoculated with the mold *Aspergillus oryzae*.

konbu (*kombu*) – a large brown kelp (*Saccharina japonica*).

kona wasabi – artificial *wasabi* powder made from horseradish to which green food colour and mustard powder are added.

kōnomono – formal word for *tsukemono*; originally meant a thing associated with incense.

kuchi atari – mouthfeel.

kuchikami no sake – *sake* made from cooked rice chewed in the mouth; *(kuchi, mouth)*.

kusaya – salted, fermented fish, typically mackerel.

kushi – bamboo skewers used to hold food together during preparation.

kushi-gata – see *nigiri*-zushi.

kyūri – small Japanese cucumbers which can be eaten raw or as *tsukemono*.

maccha – powdered green tea.

maguro – bluefin tuna (*Thunnus thynnus*); see also *hon-maguro* and *kihada*.

makisu – bamboo rolling mat used for making *maki*-zushi.

maki-zushi – sushi roll with a sheet of *nori* either on the outside or the inside; *(maki, to roll)*.

manaita – cutting board.

Manekineko – Japanese good luck cat.

masu – volume measure used, for example, to characterize the small cedar boxes from which one drinks *sake*.

mazemori – see *moritsuke*.

meron (*uri*) – melon.

mirin – sweet rice wine with ca. 14% alcohol; used in Japanese food preparation, but not intended to be drunk.

mirugai (*mirukui*) – geoduck (*Panope generosa*).

miso – salty paste made from fermented soybeans or grain, such as rice or barley; available in many different varieties, such as red *miso* (*aka-miso*) and white *miso* (*shiro-miso*).

miso-shiru – *miso* soup.

miso-zuke – vegetables or fish pickled in *miso* paste.

moriawase – see *moritsuke*.

moritsuke (*moriawase*) – overall term for an arrangement of food, for example, sushi and *sashimi*: *hiramori* (thick pieces of *sashimi* standing on edge), *yosemori* (two or three different pieces placed closely together to create a contrast), *mazemori* (a representative selection of *nigiri*-zushi arranged on a platter), *kasanemori* (slices placed so that they overlap), *sugimori* (pieces arranged at an angle to each other to form a slanting pile), *ōmori* (pieces placed in a pile, used for food offerings in the temples), and *chirashimori* (assorted pieces spread out with distance between them).

moto – mash for production of *sake* consisting of the fermentation medium *kōji* to which is added vinegared, cooked rice and a pure yeast culture.

mu – emptiness (*Zen* expression).

nama fu – unprocessed *fu*; (*nama*, raw).

nama-zushi – sushi with quickly fermented rice, best known from *nigiri*-zushi. In contrast to the slowly fermented *nare*-zushi, the rice in *nama*-zushi is eaten.

namida – tears; used as sushi slang for *wasabi*.

nare-zushi – the original form of sushi (aged sushi), in which slowly fermented rice serves to preserve fish, for example, carp in the form of *funa*-zushi. The rice in this type of sushi is not eaten.

nasu – small Japanese eggplant (*Solanum melongena*).

nazuke – pickling in brine (*shio-zuke*), see also *tsukemono*.

neta – things (*tane*) for putting on *nigiri*-zushi.

niban dashi – second *dashi*.

nigari – sea salt (bittern), predominantly consisting of magnesium chloride, traditionally used in Japan as a coagulant in *tofu* production.

nigiri-zushi – hand shaped sushi made of small rice balls on which things (*tane*) are placed, e.g., raw fish or shellfish (also called *edomae-zushi*). *Nigiri* means to grasp or hold tightly with the hand. There are five classical ways of forming the rice ball: *kushi-gata*, *hako-gata*, *tawara-gata*, *funa-gata*, and *ōgi-gata*, but *kushi-gata* is the most common.

nimono-dane – cooked and simmered things (*tane*) placed on sushi, e.g., octopus, some bivalves, and eel.

nishin – herring (*Clupea pallasii*).

nitsume – special sauce made from eel stock, used for glazing sushi eel (*anago* or *unagi*).

Nō – classical Japanese musical drama, characterized by the use of stereotypical masks, a slow tempo, and unadorned elegance.

nojime – fish which must be ripened before eating. The fish has gone through *rigor mortis* and has been frozen for a period of time, as a result of which its taste and texture have changed due to natural decomposition. *Nojimi* for sushi-*dane* must be eaten immediately after it has been thawed. Examples include red fish such as tuna or salmon.

nori – fronds from the red alga *Porphyra* which have been chopped, pressed, dried, and possibly toasted to make paper thin sheets used for, among other things, making *maki*-zushi. *Hoshi-nori* is dried *nori* and *yaki-nori* is roasted, dry *nori* that is often spiced with *shōyu*.

nuka – rice bran.

nuka-doko (*toku*) – fermentation medium based on rice bran, used to make *takuan-zuke* from white radishes (*daikon*) and other products.

nuka-miso – 'smelly women', an expression used in earlier times referring to Japanese housewives whose hands took on an odour from the daily stirring of the fermentation medium, *nuka-doko*.

nuka-zuke – *tsukemono* made by preserving vegetables in rice bran.

oboro (*denbu*) – chopped, cooked, and pressed fish or shellfish formed into a solid that is flavoured and coloured; primarily sold as imitation crab.

odori – 'dancing shrimp', sweet shrimp (*amaebi*) eaten while still alive and hence said to move in the mouth when eaten.

ōgi-gata – see *nigiri*-zushi.

o-hitsu – traditional Japanese wooden container for storing and serving warm, cooked rice.

ohyō – Pacific halibut (*Hippoglossus stenolepis*).

omakase – sushi meal where the chef is given discretion to select the dishes and decide on the order in which they are served.

ōmori – pieces of food arranged in a pile, e.g., sushi. In the temples, offerings to the gods of food are presented in this manner.

onigiri – a ball of sushi rice, possibly with some filling, wrapped in *nori*.

onji – a Japanese word corresponding to a sound; analogous to a syllable.

oroshi-gane – grater.

oshibako – wooden mold for making pressed sushi, *oshi*-zushi.

oshibori – damp cloth for wiping the face and hands.

oshinko – see *shinko*.

oshi-zushi – pressed sushi, typically made with mackerel (*saba*); characteristic of the Osaka region (Kansai).

raku – 'unadulterated enjoyment', traditional Japanese method of making rustic ceramics containing a large proportion of sand or crushed fired clay; after firing they are cooled quickly in cold air, paper, or sawdust.

ryōri – something related to cooking or food.

saba – mackerel.

sabi – aesthetic expression for the wistful beauty found in old, worn, and dilapidated things.

sakazuki – small bowls for drinking *sake*.

sake – salmon.

sake – rice wine.

sakura niku – 'cherry coloured meat', horse meat, also called *uma*.

sanmai oroshi – three-part filleting of fish into two fillets and the skeleton; used for small rounded fish such as salmon, mackerel, and herring.

sasa-giri (*sasaberi*) – 'lace border', elaborately cut out bamboo leaves used to decorate a dish, e.g., an arrangement of sushi.

sashimi – sliced raw fish or shellfish.

sencha – good quality Japanese green tea.

sengiri – sliced into thin strips, julienned.

shabu-shabu – onomatopoetic expression for a dish in which thin pieces of meat and vegetables are quickly cooked in a pot with soup stock (*dashi*).

shamoji – wooden spatula (also called *ki-jakushi*).

shari – sushi bar slang for sushi rice.

shiba-zuke – eggplant pickled in brine (*tsuke-mono*), usually with red *shiso*.

shiitake – the mushroom *Lentinus edodes*.

shimafuri – blanching, e.g., of fish skin.

shinko (*oshinko*) – vegetables pickled in brine, often ones which are lightly pickled and not fully preserved. *Oshinko* literally means 'new fragrance'.

Shinto – Japanese religion based on nature worship.

shio – salt.

shio-zuke – pickling in brine (*nazuke*).

shirako – fish sperm sac.

shiromi – white things (*tane*) placed on sushi, e.g., fish with white muscle meat.

shiro-miso – white *miso*.

shiru – soup.

shiso – leaf mint (*Perilla frutecens*) found in red (*aka-jiso*), green (*ao-jiso*), and green-red (*aoaka-jiso*) varieties.

shitazawari – tonguefeel.

shōchū – distilled rice brandy with 36-45% alcohol content.

shōga – ginger (*Zingiber officinale*).

shōjin ryōri – classical vegetarian temple food prepared in accordance with Buddhist directives; introduced in Japanese temples in the 6th Century and became more widespread in the 13th Century after *Zen* gained prominence; consists of food prepared from soybeans (*tofu*, *miso*, *shōyu*), mushrooms, seaweed, and *fu*.

shōyu – soy sauce.

su – rice vinegar containing about ca. 4% acetic acid.

sudare – bamboo mat, for example, a *makisu* for rolling *maki*-zushi.

sugata-zushi – whole fish stuffed with sushi rice, cut up, and served in the original form.

suigimori – see *moritsuke*.

suihanki – automatic rice cooker, usually electric.

suimono – clear soup made from the first *dashi* (*ichiban dashi*).

sujime – curing technique involving salt and vinegar; used, for example, on oily fish like mackerel where it mellows the flavour and firms the texture.

sushi-dane – see *tane*.

sushi-meshi – rice made ready for sushi.

sushinoko (*sushi-ko*) – powder additive for sushi rice consisting of dehydrated rice vinegar with salt and sugar.

sushiya – word used for sushi bar, or sushi restaurant, or sushi chef.

suzuki – Japanese sea bass.

tai (*ma-dai*)– red seabream.

takara-bune – little wooden boat used for presentation of sushi and *sashimi*. The original meaning is related to the treasure ships which transported valuable cargo from China to Japan.

tako – octopus (*Octopus vulgaris*).

tako-biki – 'octopus cutter', *sashimi* knife with a blunt tip used in Tokyo and eastern Japan.

takuan-zuke – pickled, white radish (*daikon*), a form of *nuka-zuke*.

tamago – egg, usually a chicken egg, but can also be a quail egg.

tamago-yaki – rolled omelette made from eggs (*tamago*); prepared in a special rectangular pan (*tamago-yaki-nabe*).

tamari – soy sauce made without wheat.

tane (*neta*) – expression for the individual pieces of topping, e.g., of fish or shellfish, placed on *nigiri*-zushi. *Tane* becomes *dane* when it follows another word.

tataki – method of preparing a fish fillet, searing it very lightly on all sides and then slicing it like *sashimi*. *Tataki* actually means to hit or break into pieces and alludes to the crushed ginger which is often placed on the grilled fish.

tatami – floor mat made of woven rice straw.

tawara-gata – see *nigiri*-zushi.

tazuna-zushi – multi-coloured (usually red, green, and white) inside-out *maki* roll, where fish and vegetables (either cucumber or avocado) create a special rainbow effect on the outside.

tekka-maki – *maki*-zushi with tuna. *Tekka* means red-hot iron and refers to the red colour of the tuna.

temaki-zushi – hand rolled sushi, for example, in a cone.

temari-zushi – small hand shaped balls of sushi rice with fish or *shiso* leaves; pressed together using transparent kitchen wrap.

tempura (*tenpura*) – deep-fried fish, shellfish, or vegetables.

tobiko (*tobiuonoko*) – flying fish (*tobiuo*) roe.

tofu – coagulated, protein-rich solid made from soy milk.

toishi – whetting stone.

toku – see *nuka-doko*.

toro – 'to melt', the sought-after fatty meat from the tuna belly.

tsukemono – different ways of pickling and preserving primarily vegetables, but also fruits. *Shio-zuke* (*nazuke*) is light brine pickling of cucumbers and eggplants, as well as Japanese apricots or plums (*umeboshi*). *Su-zuke* is pickling in rice vinegar. *Nuke-zuke* is pickling in a fermentation medium made from rice bran, for example, *takuan-zuke* made with white radish (*daikon*). *Miso-zuke* utilizes *miso* mixed with sake as a fermentation medium. *Kōji-zuke* uses *kōji*, which consists of rice bran to which the yeast, *Aspergillus oryzae*, is added as the medium.

Tsukiji – the fish market in Tokyo.

uchiwa – fan made with a frame of split bamboo reeds on which is glued paper or silk; used to cool sushi rice.

uma – raw horse meat (also *sakura niku*, 'cherry red meat'). *Sashimi* made with *uma* is also called *basashi*.

umami – 'the fifth taste' or 'meat taste', especially brought out by *monosodium glutamate* ('the third spice') and associated with the taste of such foods as *konbu*, *shiitake*, and *katsuobushi*.

ume – Japanese apricot that resembles a plum.

umeboshi – dried and brine-pickled Japanese apricots (*ume*) or plums.

unagi – freshwater eel that lives in the rivers and lakes in Japan.

unagiba-bōchō – sushi and sashimi knife for cutting trimmed fish and shell fish.

uni – sea urchin.

uramaki – inside out *maki*-roll that has the sheet of *nori* on the inside and the rice on the outside.

usuba-bōchō – heavy knife with a wide blade and even edge for peeling and cutting vegetables.

wabi – complex aesthetic expression used to describe a person, an object, or a living thing characterized by modesty, humility, aloneness, wistfulness, simplicity, or stillness.

wakame – dark green edible kelp with ribbon-like fronds and a mild *umami* taste.

waribashi – disposable wood or bamboo chopsticks.

wasabi – Japanese horseradish (*Wasabia japonica*).

yaki-nori (*ajitsuke nori*) – toasted *nori* sheets, flavoured with *shōyu* or sesame oil and often used as a topping (*furikake*) sprinkled on rice and in salads.

yanagiba-bōchō – the classical Japanese *sashimi* and sushi knife with a narrow blade and an almost even edge.

yōkan – Japanese confectionery or candy based on red *azuki* bean paste made into a solid jelly using sugar and thickened with agar (*kanten*); called *yōkan cha* when green tea is added to it.

yonezu – vinegar made exclusively with rice.

yosemori – see *moritsuke*.

yukari furikake – type of *furikake* consisting of roasted, crushed red *shiso* mixed with salt.

yukinoshita – *enokitake* mushrooms (*Flammulina velutipes*).

zaru – bamboo sieve.

Zen – Japanese-Chinese meditative school of Buddhism with philosophical overtones.

$$f(z, z_i, z_j) \equiv \frac{1}{z_j - z_i} \int_{z_i}^{z_j} \theta(z' - z)\theta(z + \Delta z -$$

$$\mathbf{F}_{ij}^D = -\gamma_{ij}(1 - r_{ij}/r_0)^2 (\hat{\mathbf{r}}_{ij} \cdot \mathbf{v}_{ij})\hat{\mathbf{r}}_{ij}; \text{ for } r$$

$$\mathrm{i}\mathcal{L}_r = \sum_{j=1}^{f} \dot{x}_j \frac{\partial}{\partial x_j}; \quad \mathrm{i}\mathcal{L}_p = \sum_{j=1}^{f} F_j \frac{\partial}{\partial p_j}; \quad \mathrm{e}^{\mathrm{i}(\mathcal{L}_1 + \mathcal{L}}$$

$$\mathrm{e}^{\mathrm{i}\mathcal{L}_\alpha \tau} \simeq \mathrm{e}^{\mathrm{i}\mathcal{L}_{\alpha,p}(\tau/2)} \mathrm{e}^{\mathrm{i}\mathcal{L}_{\alpha,r}\tau} \mathrm{e}^{\mathrm{i}\mathcal{L}_{\alpha,p}(\tau/2)}; \quad \dot{\mathbf{p}}_i = \mathbf{F}_i^C$$

SCIENTIFIC TERMINOLOGY

AA – see *arachidonic acid*.

acetic acid – (vinegar) organic acid formed by bacterial and fungal fermentation of *sugars*.

acid – large class of chemical compounds that release hydrogen ions when dissolved in water. Acids generally have a sour taste and can be neutralized by bases. Examples are acetic acid, citric acid, lactic acid, fatty acids, and amino acids.

acidity – see *pH*.

actin – *protein* molecules and thin filaments made of it that create structure inside the *cell*, on the surface of the cell and, for example, in muscles. Individual actin molecules can *polymerize* into long filaments with a thickness of only seven nanometers but a length of up to several micrometers. *Crosslinkages* among the actin filaments form a network which helps to give the cell shape. In the muscles, long actin filaments act somewhat like tracks along which the *molecular motor myosin* can slide during muscle contraction.

adenosine triphosphate – (ATP) chemical compound (*nucleotide*) that is a source of energy; together with ADP (adenosine diphosphate) it is involved in virtually all biochemical processes that require energy.

agar – a mixture of *polysaccharides* extracted from red seaweed; used as a thickening agent.

alcohol – generic term for a large group of chemical substances which contain an –OH group. 'Ordinary alcohol' is *ethanol. Cholesterol* is also an alcohol.

aldehydes – together with *ketones* and *esters*, aldehydes make up the chemical compounds known as the carbonyl compounds, which is to say that they contain the group –C=O.

alkaloids – nitrogen containing basic group of chemical compounds that include, among other substances, *caffeine* and nicotine; many are poisonous.

alpha-linolenic acid – polyunsaturated *omega-3 fatty acid* with 18 carbon atoms and three double bonds, $(18:3)(9,12,15)$ $CH_3-CH_2-CH=CH-CH_2-CH=CH-CH_2-CH=CH-(CH_2)_7-COOH$. It is the starting point for the formation of superunsaturated omega-3 fatty acids, e.g., DHA (*docosahexaenoic acid*) and EPA (*eicosapentaenoic acid*).

amines – substances which contain nitrogen, for example, an amino group, $-NH_2$, in the primary amines.

amino acids – small molecules made up of between 10 and 40 atoms, which in addition to carbon, hydrogen, and oxygen always contain an amino group – NH_2. Amino acids are the fundamental building blocks of *proteins*. Examples include *glycine, glutamic acid*, alanine, proline, and arginine. Nature makes use of 20 different, specific amino acids to construct proteins, which are chains of amino acids bound together with so-called peptide bonds. Short chains are called polypeptides and long ones, proteins. In food, amino acids are often found bound together in proteins and also as free amino acids which can have an affect on taste. An example is *glutamic acid* which is the basis of the *umami* taste.

Of the 20 natural amino acids, there are 9 essential ones that our bodies cannot themselves produce and which we must get from food (valine, leucine, lysine, histidine, isoleucine, methionine, phenylalanine, threonine, and tryptophan).

amphiphile – a substance or molecule with mixed feelings toward water. Typically used to describe molecules, such as *proteins* and *lipids*, which consist of two parts, one of which attracts water and the other which repels it.

amylopectin – *polysaccharide* consisting of a branched network of *glucose* molecules; together with *amylose* it is the most important ingredient in starch.

amylose – *polysaccharide* consisting of long, linear chains of *glucose* molecules; together with *amylopectin* it is the most important ingredient in starch.

anisakis – herring worm (*Anisakis simplex*), parasitic nematode sometimes found in mackerel, herring, cod, and squid.

anthocyanin – red pigment in plants such as red *shiso*. Because anthocyanins dissolve readily in water, they can easily be used as a dye. The colour is very sensitive to acidity and contact with metal. Low *pH* conserves the red colour, but metal ions can cause the colour to shift to blue or green.

antibiotics – substances that fight microorganisms such as bacteria and fungi. Penicillin is an antibiotic.

antioxidant – substance that prevents *oxidation* of other substances, for example, unsaturated *fats* which oxidize easily (become rancid). *Ascorbic acid* (vitamin C), vitamin E, and green chlorophyll are important antioxidants in foodstuffs.

arachidonic acid – (AA) superunsaturated long-chain *fatty acids* with 20 carbon atoms and four double bonds, (20:4)(5,8,11,14) CH_3–$CH=CH$–CH_2–$CH=CH$–CH_2–$CH=CH$–CH_2–$CH=CH$–$(CH_2)_7$–$COOH$; belongs to the *omega-6* family.

ascorbic acid – *vitamin* C.

astaxanthin – orange-red pigment e.g., found in fish and shellfish. Astaxanthin is a *carotenoid* and is chemically related to the pigment what gives carrots their characteristic yellow-orange colour. In intact shells of crustaceans, the astaxanthin is bound in a *protein* complex (*crustacyanin*), in which form it is not red, but blueish green or reddish brown.

astringency – a harsh, biting physical sensation in the mouth, which is not a true taste sensation; well known from black tea that has steeped too long or from red wine that contains large quantities of *tannic acid* (*tannins*, *phenols*). Astringency is experienced because the tannins bind with the *proteins* in the saliva, causing a dry, chafing feeling, which can be perceived as either pleasant or unpleasant depending on context.

atom – the fundamental, smallest particle of an element, for example, hydrogen (H), oxygen (O), nitrogen (N), carbon (C), and sulphur (S). *Molecules* are made up of atoms. held together by chemical bonds.

ATP – see *adenosine triphosphate*.

bittern – see *nigari*.

bivalves – molluscs having a shell consisting of two hinged valves, e.g., oysters and clams.

bromophenols – bromine containing *phenol* compounds, stored in, for example, saltwater fish and the algae on which fish or their prey live. Bromophenols have the smell we associate with a fresh sea breeze.

caffeine – *alkaloid*, stimulant, found in coffee and tea, as well as in other foods.

calcite – see *calcium carbonate*.

calcium carbonate – limestone, $CaCO_3$.

calcium sulphate – gypsum, $CaSO_4$.

canthaxanthin – industrially produced *carotenoid*, sometimes added to fish fodder so that the fish muscles turn red. Substitute for natural *astaxanthin* on fish farms.

capsaicin – organic substance responsible for the strong taste of chili peppers.

carbohydrates – *saccharides* or sugars, a large group of chemical compounds which primarily consist of oxygen, hydrogen, and carbon. The simple saccharides, *monosaccharides* and disaccharides, are sweet and include the ordinary sugars, such as glucose, fructose, and galactose, as well as sucrose, lactose, and maltose. *Starch, cellulose,* and *glycogen* are polysaccharides. Carbohydrates are formed in plants and algae by photosynthesis in which carbon dioxide and water combine. Carbohydrates make up the fuel for the metabolism of all animals.

carotenoid – group of red-orange pigments in plants and animals, e.g., *astaxanthin* in shellfish and carotene in carrots.

casein – milk *protein*; when milk is acidified, the casein undergoes a *coagulation process* to form cheese curds.

cell – the smallest living entity of an organism; protected from its surroundings by a *cell membrane*, which is a part of the cell wall. Some organisms are unicellular, e.g., bacteria and yeast. Others are multicellular with a few hundred cells to billions of them. A human has about 100,000 billion cells.

cellulose – *polysaccharide* built up of linear chains of *glucose* but, in contrast to *starch*, the glucose chains are bound together closely in such a way that cellulose is not water soluble, nor can our stomachs digest it.

cephalopods – molluscs with a reduced outer or inner shell, or no shell at all, for example, octopuses and squid; said to be free swimming because they use their arms to move.

cholesterol – fat found in large quantities in all animal *cell membranes*; basis for the formation of sex hormones, *vitamin* D, and bile salts. The distribution and transportation of cholesterol in the body is mediated by certain *lipoproteins*. If there is an imbalance between this transportation system and the liver's capacity to create and break down cholesterol, the danger of atherosclerosis increases. Cephalopods and crustaceans contain a fair amount of cholesterol, especially cuttlefish. Great quantities of cholesterol are also found in fish roe.

citric acid – organic acid which imparts the characteristic sour taste to citrus fruit.

coagulation – process by which something clots together (coagulates), e.g., blood proteins which form a blood clot or milk proteins (*casein*) which form cheese curds.

collagen – the most important *protein* in connective tissue, where it forms stiff fibres that hold the muscle fibres together and bind them to skin and joints. Collagen consists of several protein molecules that are twisted together in threes in a triple spiral (helix) in the same way as a rope. On being warmed, this spiral is dissolved, loses its stiffness, and becomes *gelatine*.

colloidal particles – particles which are so small that they can remain in suspension in a liquid, e.g., fat particles in homogenized milk or clay particles in a glacial lake.

convection – circulatory transport motion in a liquid or gas caused by, for example, temperature differences.

crosslinking – formation of chemical bonds across and between long-chained *polymers*, for example, *proteins*. Crosslinking of fibres and polymers is a way to make soft materials more robust and tough. Industrially, crosslinking is used in such processes as vulcanization which converts the polymer polyisoprene to rubber. This is what imparts strength and good elastic properties to rubber.

crustacyanin – blue-green or red-brown *protein* complex found, for example, in the shells of crustaceans where it is bound to the orange-red substance *astaxanthin*. When it is broken down by heating or digestive processes, crustacyanin *denatures* and the red colour of astaxanthin becomes pronounced.

decapods – crustaceans with ten legs, for example, shrimp and lobsters.

denaturing – word often used to describe the process which the proteins undergo when they are heated or affected by salt or acid (e.g., when marinated or pickled).

deoxyribonucleic acid – (DNA) *polynucleotide* consisting of a chain of nucleic acids together with *sugars* and phosphate groups; basis for the genetic information encoded in genetic material and the genome. In the genome, DNA forms a double helix in which two DNA chains spiral around each other.

dermis – the innermost layer of the skin, consisting of living cells complete with blood supply and nerve endings. The *epidermis* lies on top of the dermis.

DHA – see *docosahexaenoic acid.*

dimethyl sulphide – sulphur compound (CH_3SCH_3) with a characteristic strong odour emitted by, for example, heated milk, cooked mussels, and rotten seaweed.

dioxin – common designation for a group of organic compounds that contain chlorine, are soluble in fats, and which accumulate in the fatty tissue of animals. Garbage incineration, steel and pesticide manufacture, and forest fires are the principal sources of dioxin in the environment. It is potentially threatening to health even in minute quantities.

DNA – see *deoxyribonucleic acid.*

docosahexaenoic acid – (DHA) superunsaturated, long chain fatty acid with 22 carbon atoms and six double bonds; member of the *omega*-3 family.

echinoderms – phylum of marine invertebrates including, e.g., sea urchins, starfish, and sea cucumbers.

eicosanoids – hormones and signaling molecules formed from *omega*-3 or *omega*-6 *fatty acids* that are important for regulation of such things as blood flow and the immune defences.

eicosapentaenoic acid – (EPA) superunsaturated, long chain fatty acid with 20 carbon atoms and five double bonds; member of the *omega*-3 family.

emulsion – mixture consisting of an oil-like substance, for example, a fat, dispersed in small droplets in another liquid in which it is only sparingly soluble, e.g., oil in vinegar.

Mayonnaise and ice cream are examples of emulsions. Emulsification can be enhanced with *emulsifiers*, substances that can bind oil and liquid together, e.g., *amphiphiles* such as *lipids*. Emulsifiers lower the *surface tension* between the oil and the liquid.

enzyme – *protein* that functions as a catalyst for a chemical or biochemical reaction.

EPA – see *eicosapentaenoic acid.*

epidermis – outer layer of the skin that lies closest to the surface on top of the *dermis*. The layer consists of dead cells in a dense structure of *proteins* and *fats* and is responsible for the skin's exceptional properties as a protective barrier.

ester – chemical compound resulting from the reaction of an *acid* with an *alcohol*. Together with *ketones* and *aldehydes*, esters constitute what is known as the carbonyl compounds (contain the –C=O group) and form the most important taste substances, for example, those formed in the course of fermentation processes.

ethanol – 'ordinary' *alcohol*, CH_3–CH_2–OH.

ethylene – gas, CH_2=CH_2, which acts as a natural ripening agent for fruits.

eukaryote – higher organism, either unicellular or multicellular, whose genetic material is enclosed in a nucleus. Fungi, plants, seaweed, and animals are eukaryotes. Primitive unicellular organisms which lack a nucleus are called *prokaryotes*. All bacteria are prokaryotes.

fat – common designation for an extensive class of substances that are not soluble in water. Fats can be solid, e.g., butter and wax, or liquid, e.g., olive oil and fish oil. The melting point of a fat has major significance for its taste and nutritional value. A typical fat consists of a long chain of carbon atoms, which can be either saturated or unsaturated. An important type of naturally occurring fats are *lipids*, which are composed of *fatty acids* bound to a variety of other substances, for example, *amino acids* and *saccharides*. Lipids are *amphiphilic* molecules.

fatty acid – a compound consisting of a long chain of carbon atoms with a carboxylic acid group. Adjoining atoms in the chain are chemically joined by either a single or double bond. Those with the most double bonds are described as the most unsaturated. If only single bonds are present the fatty acid is said to be fully saturated. Monounsaturated fatty acids have a single double bond, e.g., *oleic acid* from olive oil. Polyunsaturated fatty acids have more than one double bond, e.g., two double bonds in *linoleic acid* from soybeans or three double bonds in *alpha-linolenic acid* found in flax seed and seaweed. Super-unsaturated fatty acids have more than four double bonds, e.g., six double bonds in DHA (*docosahexaenoic acid*) derived from fish oil. Essential fatty acids are fatty acids that the human body cannot itself produce and which, therefore, have to be obtained from food sources. There are two families of these, both polyunsaturated fatty acids: *linoleic acid* and *alpha-linolenic acid*. They are the progenitors of two important types of fatty acids, the *omega*-3 and *omega*-6 fatty acids.

fermentation – process in which microorganisms (or microbes) such as yeast or bacteria convert *sugars* to *alcohol*, e.g., *ethanol*, or to acid, e.g., *vinegar*.

flavones – together with flavonols and flavenes constitute a particular group of *phenols* found in plants and fruits that help to impart a bitter and *astringent* taste, for example, that found in green tea and in citrus fruits.

gastrophysics – qualitative reflections on, and quantitative examinations of, foods, their handling, conversion, and processing, focusing on physical effects and explanations.

gastropods – molluscs with a single, usually coiled shell, e.g., snails.

gel – technical term for a network of molecules that contain large quantities of water but are also somewhat stiff like a solid; formed by *gelation* processes, for example, when egg whites are heated or *gelatine* is cooled.

gelatine – the same *protein* as the one found in the form of *collagen* in connective tissue. In contrast to collagen, gelatine is soluble in water and is formed when collagen is heated, dissolving the stiff fibres therein. On cooling, the stiff fibre structure of collagen is not formed again; in its stead a *gel* containing water is produced, a process called *gelation*.

gelation – see *gelatine*.

gene – a sequence of *nucleotides* of DNA that, among its other functions, contains the genetic information of an organism (hereditary material).

genome – the combined genetic information of a given organism, namely, all the genes.

gingerol – organic substance which imparts the sharp taste in ginger; in the same chemical family as the strong taste substances *piperin* and *capsaicin* in black pepper and chili.

gliadin – a *protein* in *gluten*.

glucose – sugar or *monosaccharide*, $C_6H_{12}O_6$, that is the most important *carbohydrate* in plants and animals. In plants glucose is formed by photosynthesis.

glucosinolate – class of organic compounds that contain sulphur, nitrogen, and a sugar group (*glucose*), e.g., *sinigrin* in black mustard, cabbage, horseradish, and *wasabi*. When water is present, the glucosinolates are converted, with the help of the *enzyme myrosinase*, to *isothiocyanate* which has an unpleasant smell and a sharp, irritating taste. In this way, plants make use of glucosinolates as a natural means of defence.

glutamic acid – *amino acid* found in such foods as fish, shellfish, and seaweed, often in the form of a salt, *monosodium glutamate*, which is the basis for the *umami* taste.

gluten – certain *proteins* (especially *gliadin* and *glutenin*), found in wheat, which enhance the baking properties of dough made with wheat flour. Kneading stretches the proteins and forms an elastic, water-binding network that traps the bubbles of carbon dioxide which are formed when the dough rises.

glutenin – a *protein* in *gluten*.

glycine – the smallest and simplest *amino acid*, $CH_2(NH_2)–COOH$.

glycogen – branched *polysaccharide* molecule consisting of *glucose* units. Glycogen acts as an energy storage depot in the liver and white musculature of fish and shellfish.

GMP – see *guanosine monophosphate*.

guanine – basic component in the formation of *nucleic acid*. Guanine crystals are found in the skin of some fish which live near the surface of the sea, e.g., herring and mackerel. These crystals impart a silvery-white sheen to the skin of these fish.

guanosine monophosphate – (GMP) *nucleotide* formed together with *inosine monophosphate* (*IMP*) when the energy storing biomolecule *ATP* is broken down by the cells to produce energy. This substance has *umami* taste and is 10 to 20 times more potent than *MSG*.

hemoglobin – reddish, iron containing *protein* that can bind oxygen and is the basis for the blood's ability to transport oxygen within the body.

hydrocarbons – organic compounds that contain carbon and hydrogen, for example, in the form of a chain of carbon atoms in *oils* or *fats*.

hydrogen bonding – a particular form of polar, chemical bonding, based on the special ability of the hydrogen atom to donate an electron to another suitable atom, such as oxygen. Hydrogen bonds are extensive in *water* and contribute to its singular properties with regard to melting and boiling point, specific heat, etc. Each water molecule can form up to four hydrogen bonds with water or other types of molecules that also have the capacity to form hydrogen bonds. Hydrogen bonds are important for the formation of stable structures in intact *proteins* and *enzymes*.

hydrophilic – loves water; typically characterizes a molecule that is soluble in water but not in oil.

hydrophobic – avoids water; typically characterizes a molecule that is not water soluble, but dissolves easily in oil.

IMP – see *inosine monophosphate*.

inosine monophosphate – (IMP) *nucleotide* formed together with *guanosine monophosphate* (GMP) when the energy storing biomolecule *ATP* is broken down by the cells to produce energy; has *umami* taste and is 10 to 20 times more potent than *MSG*.

ion – electrically charged atom or molecule.

isothiocyanates – chemical compounds with the S=C=N–group. These substances are malodourous and are formed, for example, when mustard seeds, cabbage, horseradish, or *wasabi* are crushed. All of these contain *glucosinolates*, such as *sinigrin*, which, after the cells are destroyed by the mechanical action of grating or chopping, are converted to isothiocyanates in the presence of water and with the assistance of *thioglucosidase* enzymes. The release of isothiocyanates forms a part of the plants' own chemical defence system.

ketones – together with *aldehydes* and *esters* make up the chemical compounds called carbonyl compounds, which is to say that they contain the group –C=O. Many taste substances which are formed during *fermentation* are ketones.

lactic acid – simple organic acid, $CH_3–CH(OH)–COOH$ which is produced, e.g., by lactic acid bacteria. It is also formed in the muscles when *glycogen* is consumed in the presence of oxygen.

lanosterol – a primitive *sterol* that is the chemical precursor of *cholesterol*.

lenthionine – a cyclical organic molecule that contains carbon and sulphur. The special aroma of *shiitake* mushrooms (*Lentinus edodes*) is due to the *enzymatic* formation of lenthionine. Lenthionine possibly suppresses the formation of carcinogenic nitrosamines in the digestive system.

lentinan – a polysaccharide found in *shiitake* mushrooms; helps to boost the immune system and may be effective as an anti-cancer agent.

limonene – chemical substance belonging to the group of *terpenes* known from the aroma of citrus rinds, dill, pepper, and caraway seed; is of the same family as perillic acid found in *shiso*.

linoleic acid – polyunsaturated *omega-6 fatty acid* with 18 carbon atoms and two double bonds, $(18:2)(9,12)$ $CH_3–(CH_2)_4–CH=CH–CH_2–CH=CH–(CH_2)_7–COOH$; basis for the formation of superunsaturated fatty acids of the omega-6 family, e.g., *arachidonic acid*.

linolenic acid – see *alpha-linolenic acid*.

lipid – *amphiphilic fat* that consists of a water soluble part and an oil soluble part.

lipid membrane – double layer of *lipid molecules* with water on both sides of it.

lipoprotein – complex of *fats* (*lipids*) and *proteins*. Lipoproteins are important for the transport of fats, e.g., *cholesterol*, in the body via the bloodstream.

liposome – a closed shell consisting of a double layer of *lipids* with water on both sides.

lipoxygenase – *enzyme*. In fish, for example, it can oxidate unsaturated *fats*, among them *linoleic acid* and *alpha-linolenic acid*, generating a volatile aromatic substance which is also associated with the odour of plants.

macromolecule – large molecule, e.g., a *protein* or DNA.

macromolecular assembly – assembly of large molecules, e.g., *lipids* organized in a *membrane*.

magnesium chloride – $MgCl_2$; among other uses it serves as a coagulant in the production of *tofu* from soy milk.

Maillard reactions – class of chemical reactions which are typically associated with non-*enzymatic* browning occurring, for example, during frying, grilling, or baking. In the course of these reactions, *carbohydrates* bind with *amino acids* from *proteins* and, after a series of intermediate steps, form a series of poorly characterized brown pigments and aromatic substances collectively known as *melanoids*. These substances give rise to a broad spectrum of taste and smell sensations ranging from the flower- and plant-like to meat- and earth-like.

mannitol – sugar *alcohol* found, e.g., in mushrooms and seaweed to which it imparts its characteristic sweet taste; ensures that the seaweed maintains the correct *osmotic* balance in salt water. As it cannot be converted in the body, it has few calories.

melanins – see *melanoids*.

melanoids – and *melanins* are brown, aromatic pigments that formed as compounds of *carbohydrates* and *amino acids*, for example, in the course of *Maillard reactions* (browning). Melanoids are also formed in the *fermentation* of soybeans to produce soy sauce (*shōyu*). *Melanin* is the black-brown pigment in black roe (caviar) from sturgeon.

membrane – the boundary between a *cell* and its surroundings (cell wall). This term is used particularly to refer to the double layer of *lipids* (*fats*) which form the middle part of the cell wall.

metabolic syndrome – composite of life style dependent, non-communicable diseases which are attributed to diet, especially cardiovascular disorders, obesity, type 2 diabetes, high blood pressure, and possibly psychiatric disorders.

methylmercury – abbreviation for monomethylmercury, most often found as the *ion* CH_3- Hg^+. Methylmercury appears as an environmental toxin, for example, in fish.

mitochondria – organelles in *cells* which produce energy in the form of *ATP*.

molecule – assembly of two or more *atoms* which are held together by chemical bonds, e.g., water (H_2O), which consists of two hydrogen atoms (hydrogen, H) and one oxygen atom (oxygen, O).

molecular gastronomy – study of the properties, at the molecular level, of food ingredients, as well as their interdependent relationships and the changes they undergo during preparation and consumption.

molecular motor – *macromolecule*, typically a *protein*, which carries out a mechanical function on a molecular level. For example, *myosin* in muscle connective tissue is a molecular motor that can slide over *actin* molecules and in this way cause the muscle to contract. Other molecular motors execute a rotating motion when *ATP* is formed or help to pull the two parts of the cell nucleus apart during cell division. It is a molecular motor, kinesin, which makes the flagella on a microorganism rotate and, in this way, propel the organism forward.

molluscs – a phylum of invertebrate animals. Most have an exterior skeleton, like mussels, oysters, and snails, or are *cephalopods* with a reduced outer shell, an internal shell, or no shell at all (e.g., octopuses and cuttlefish).

monosodium glutamate – (MSG) sodium salt of the *amino acid glutamic acid*, also known as 'the third spice' because it is the one most widely used after salt and pepper; imparts the *umami* taste.

MSG – see *monosodium glutamate*.

mustard oil – see *isothiocyanates*.

mycelium – branched filament which makes up the root mat of a mushroom.

myocommata – see *myotomes*.

myoglobin – red *protein* containing iron found in muscles where it transports oxygen from the blood to the muscle fibres. It is myoglobin which gives some animal and fish muscles their red colour. On heating, the myoglobin *denatures* and takes on a brownish colour.

myosepta – see *myotomes*.

myosin – *protein* that functions as a *molecular motor* in muscle connective tissue where it slides along the *actin* fibres.

myotomes – thin layers of muscle fibre in fish. The layers are typically from a few millimetres to one centimetre thick. These layer divisions are recognizable in the flakes into which a cooked fish separates. Myotomes are held together by some fragile layers of connective tissue, namely, *myosepta* along the fibre bundles and *myocommata* across the fibre bundles. Myocommata extend from the innermost layer of skin (*dermis*) of the fish to the bone and are arranged in a zigzag formation.

myrosinase – *enzyme* of the *thioglucosidase* type which converts *glucosinolates* to *isothiocyanates*.

nucleic acid – chemical designation for a macromolecule made up of *nucleotides* bonded together. Nucleic acids are the building blocks of DNA, RNA, and *genomes*.

nucleotide – substance composed of a nitrogenous nucleobase (adenine, guanine, cytosine, uracil, or thymine), a *sugar*, and one or more phosphate groups. The *umami* taste substances GMP and IMP are nucleotides.

octenol – short-chain alcohol derived from the *enzymatic* breakdown of the superunsaturated *fatty acid, linoleic acid*. The aroma of freshly harvested mushrooms is due to octenol, which is formed when the cells of the mushroom are damaged, especially its lamellae (the gills under the cap). For this reason mushrooms that are not yet fully developed have a blander taste than those with mature lamellae. In addition, brown mushrooms are tastier than white ones.

oil – chemical compound containing carbon; not soluble in water; examples include, *hydrocarbons, fatty acids* and *lipids*.

oleic acid – monounsaturated *fatty acid* with 18 carbon atoms; main component of olive oil.

omega-3 fats – polyunsaturated *fats* derived from *alpha-linolenic acid*, e.g., DHA (*docosahexaenoic acid*) and EPA (*eicosapentaenoic acid*).

omega-6 fats – polyunsaturated *fats* derived from *linoleic acid*, e.g., AA (*arachidonic acid*).

osmosis – process of diffusion of particles and molecules across a barrier, for example, a *cell membrane*, which is permeable to water but impermeable to the other larger molecules, such as salt, *amino acids*, or *sugar*. The resulting imbalance is equilibrated when some of the water passes across to the side containing the large molecules. This rate of diffusion increases with the degree of *hydrophilicity* of these molecules. The osmotic effect is counterbalanced by a pressure, called the osmotic pressure, across the entire membrane. Osmosis is central to the ability of plants to draw water from the ground, into their root system, and up through their trunks and branches. The opposite process, known as reverse osmosis, in which pure water is drawn out of a solution, is used for purifying water.

osmotic pressure – see *osmosis*.

oxidation – removing one or more electrons from an *atom, ion*, or *molecule*. For example, the double bonds of unsaturated *fats* can be oxidized resulting in rancidity.

PCBs – polychlorinated biphenyls are a class of fat soluble organic compounds. PCBs had many industrial applications, e.g., as hydraulic fluids, lubricants, and cutting oils. Because PCBs naturally break down very slowly, they are biohazards that accumulate in the food chain and are found in the environment everywhere on the planet even though they have been banned since the 1970's. They cause certain skin diseases and possibly cancer.

perilla acid – substance classified as a *terpene*, found in such plants as red *shiso*. Other related substances are perilla alcohol and perilla aldehyde, which is the active anti-microbial ingredient in red *shiso*.

pH – quantitative measure for relative acidity. A pH reading of 7 is neutral and readings of below and above 7 correspond, respectively, to acidic and basic (alkaline) environments.

phenols – large group of acidic chemical substances derived from phenol (hydrobenzene). Found in plants, for example, in green tea to which it imparts its characteristic slightly bitter taste with hints of grass, hay, flowers, and seaweed. The principal type of *polyphenol* (epigallocatechin-gallate) in green tea can counteract cancer, especially stomach and throat cancers. It is thought that the relevant mechanism is that the polyphenol in-

hibits the growth of blood vessels in the tumor. Ocean fish accumulate *bromophenols*, the smell of which is associated with that of a fresh ocean breeze. Bromophenols are formed, for example, by marine algae which are consumed by fish or by their prey. *Oxidation* of phenol compounds in fruits and vegetables causes a brown discolouration when they are sliced, mashed, or affected mechanically by other means. Smoking of fish can also bring out phenol compounds, for example, in the preparation of *katsuobushi*.

phospholipid – *lipid* with a phosphate group polar head; an important component of *cell membranes* and fish muscle.

piperin – organic substance which imparts a strong taste to black pepper.

polyamide – *polymeric* chain of *amino acids*, as in a *protein*.

polymer – large molecule, either in the form of a chain or branched, composed of many identical or different units (monomers). An example is a *protein*, a naturally occurring form of the group known as *polyamides*. Polymers can be made by a polymerization process in which the individual monomers are bound together in a chemical reaction.

polynucleotide – chain of *nucleotides*, e.g., in DNA.

polyphenol – chemical compound containing several *phenol* groups.

polyphenol oxidases – *enzymes* found in green tea leaves where they produce a series of aroma substances and pigments based on the tea leaves' complex chemical *phenol* compounds.

polysaccharide – *sugar*, see *carbohydrates*; consists of several saccharide units, for example, the disaccharide lactose, which is the sweet substance in milk, or the polysaccharide *glycogen*, which is the energy storage depot in the liver and the white muscles of fish.

prokaryote – unicellular organism which lacks a nucleus. All bacteria are prokaryotes.

protein – *polyamide*, which is to say a long chain of *amino acids* bound together by peptide bonds. *Myoglobin*, an important protein in the muscles, is the source of the red colour of meat. *Receptors*, which capture signals in the cells and identify things such as taste and smell, are also proteins. *Enzymes* are a particular class of proteins whose function it is to ensure that chemical reactions take place under controlled circumstances. Proteins lose their functional ability (*denature*) and their physical properties change when they are heated or exposed to salt or acid (as in cooking, salting, or marinating).

protozoa – unicellular organisms with a nucleus, e.g., amoebae and flagellates.

pyrazine – nitrogen containing the cyclic compound $C_4H_4N_2$.

receptor – *protein* molecule that has a special ability to bind with a particular substance, for example, a smell or taste molecule. Receptors are found in all *membranes*, especially those of nerve cells.

ribonucleic acid – (RNA) *polynucleotide* which, like DNA, is made up of four nucleobases, but with uracil instead of thiamine and also with different sugar groups.

rigor mortis – temporary chemical change in the muscles occurring after death causing them to become stiff.

RNA – see *ribonucleic acid*.

saccharide – *sugar*, see *carbohydrates*.

sinigrin – chemical substance that belongs to the *glucosinolate* group, found in mustard, cabbage, horseradish, and *wasabi*, among others.

starch – mixture of the *polysaccharides amylose* and *amylopectin*.

sterol – cyclic carbon compound that consists of a *hydrophobic* core of four, fused rings. The so-called higher sterols are important for all advanced forms of life (*cholesterol* in animals, ergosterol in fungi and yeast, fucusterol in seaweed, and phytosterol in plants).

sugar – see *carbohydrate*.

surface tension – expression for a force, based on attraction between molecules, that tries to diminish a surface to the smallest possible area. It is this force that makes it possible to fill a glass with water to just above the rim. In general, *interfacial tension* is an expression for a similar force that tries to minimize the area of contact between joining surfaces, a typical case being that of oil and water. This force can be lessened and miscibility increased by adding a substance, which is active on the boundary surface, for example, soap or another *amphiphilic* substance such as a *lipid* or a suitable *protein*.

tannin – (*tannic acid*) common designation for *phenols*, which are bitter taste substances, found in red wine, black tea, and smoked products, among others.

taurine – *amino acid*, major constituent of bile where it functions as an *emulsifier* to bind *fats* and mediate the uptake of *lipids*, e.g., *cholesterol*; found in large amounts in seafood such as octopuses, squids, clams, and oysters. Strictly speaking, taurine is not a real amino acid since it lacks a carboxyl group.

terpenes – class of organic substances which are the primary constituents of olfactory substances in many plant oils, e.g., *perilla acid* and derivates thereof in red *shiso*. A related terpene is *limonene*, found in dill, pepper, and caraway seeds.

tetrodotoxin – neurotoxin (nerve poison) named after the pufferfish, *fugu* (*Tetraodontiformes*). The poison works by blocking the sodium channels in the *membranes* of the nerve cells. Less than one milligram, which is to say the quantity that can be placed on the tip of a needle, is sufficient to kill an adult.

theanine – *amino acid* commonly found in green tea leaves, among other sources.

thiamine – *vitamin* B_1.

thiazole – cyclic compound, C_3H_3NS, containing nitrogen and sulphur.

thioglucosidase – *enzyme* that converts *glucosinolates* to *isothiocyanates*.

toxin – poison, typically derived from a plant, fungus, or animal.

triglyceride – *fat* with three *fatty acid* groups.

trimethylamine – foul smelling organic substance (tertiary amine) produced, for example, by bacterial decomposition of *trimethylaminoxide* in dead fish. Trimethylaminoxide, which is odourless, is used by the *cells* of the fish to balance the *osmotic pressure* due to the saltiness of oceanic water. Fish from salty waters therefore contain more trimethylamine than those from sweet water.

turbulence – chaotic movements in gases or liquids, for example, in connection with boiling of water.

vacuole – empty space in a cell where it stores nutrients or waste products.

vinegar – see *acetic acid*.

viscosity – resistance to flow in a liquid; alternatively, the capacity of a liquid to resist when another substance is moving through it.

vitamin – group of different essential organic substances that the body itself can produce only in limited quantities and which, therefore, must be ingested. Examples are vitamins A, B, C, D, E, and K. Vitamin C (*ascorbic acid*) and vitamin E are also important *antioxidants* in foods.

water – essential chemical compound, H_2O, consisting of *molecules* made up of two hydrogen (H) atoms and one oxygen (O) atom.

wine vinegar – sharp tasting, acidic wine produced by allowing the alcohol in wine to *oxidize* to form *acetic acid*.

BIBLIOGRAPHY

ON SUSHI AND SUSHI RELATED FOODS

Andoh, E. *Washoku.* Ten Speed Press, Berkeley, 2005.

Ashkenazi, M. & J. Jacob. *The Essence of Japanese Cuisine. An Essay on Food and Culture.* University of Pennsylvania Press, Hampden Station, Baltimore, 2000.

Barber, K. *Sushi. Taste and Technique.* DK Publishing, London, 2002.

Chu, C. *The Search for Sushi. A Gastronomic Guide.* Crossbridge Publ. Co., Manhattan Beach, 2006.

Corson, T. *The Zen of Fish.* HarperCollinsPublishers, New York, 2007.

Dekura, H. *The Fine Art of Japanese Cooking.* Bay Books, Sydney, 1993.

Dekura, H. *Contemporary Japanese Cuisine.* Weatherhill Inc., Trumbull, 2001.

Dekura, H. *Sushi Modern.* Periplus Ed., Boston, 2002.

Dekura, H., B. Treloar & R. Yoshii. *The Complete Book of Sushi.* Periplus Ed., Singapore, 2004.

Detrick, M. *Sushi.* Chronicle Books, San Francisco, 1981.

Egan, A. *Edamame*. Rodale Inc., Emmaus, USA, 2003.

Fujii, M. *The Enlightened Kitchen*. Kodanska International Ltd., Tokyo, 2005.

Gusman, J. *Vegetables From the Sea*. William Morrow, HarperCollinsPublishers, New York, 2003.

Gustafson, H. *The Green Tea User's Manual*. Clarkson Potter Publ., New York, 2001.

Hosking, R. *A Dictionary of Japanese Food. Ingredients and Culture*. Tuttle Publ., Boston, 1996.

Hosking, R. *At the Japanese Table*. Oxford University Press, Oxford, 2000.

Issenberg, S. *The Sushi Economy. Globalization and the Making of a Modern Delicacy*. Gotham Books, New York, 2007.

Kawasumi, K. *The Encyclopedia of Sushi Rolls*. Graph-ha, Ltd., Tokyo, 2001.

Kazuko, E. *EasySushi*. Lothian Books, Port Melbourne, 2000.

Kazuko, E. *Japanese Food and Cooking*. Lorenz Books, London, 2001.

Kazuko, E. *New Sushi*. Jacquie Small, London, 2006.

Klippensteen, K. *Cool Tools*. Kodanska International Ltd., Tokyo, 2006.

Liley, V. *Sushi*. Whitecap, North Vancouver, 2005.

Lowry, D. *The Connoisseur's Guide to Sushi*. The Harvard Common Press, Cambridge, 2005.

Masui, K. & C. Masui. *Sushi Secrets*. Hachette, London, 2004.

Matsushisa, N. *Nobo. The Cookbook*. Kodanska International Ltd., Tokyo, 2001.

Murata, Y. *Kaiseki. The Exquisite Cuisine of Kyoto's Kikunoi Restaurant*. Kodansha International Ltd., Tokyo, 2006.

Ogawa, S. *Easy Japanese Pickling in Five Minutes to One Day*. Graph-Sha Ltd., Tokyo, 2003.

Ōmae, K. & Y. Tachibana. *The Book of Sushi*. Kodansha International Ltd., Tokyo, 1988.

Owen, S. *The Rice Book*. Frances Lincoln, London, 2003.

Sanmi, S. *The Way of Tea. A Japanese Tea Master's Almanac*. Tuttle Publishing, Boston, 2002.

Shimbo, H. *The Sushi Experience*. Alfred A. Knopf, New York, 2006.

Shimizu, K. *Tsukemono. Japanese Pickled Vegetables*. Shufunotomo Co. Ltd., Tokyo, 1993.

Strada, J. & M. T. Moreno. *Sushi for Dummies*. Wiley Publishing Inc., Indianapolis, 2004.

Sugimoto, T. & M. Iwatate. *Shunju. New Japanese Cuisine*. Periplus Ed., Singapore, 2002.

Sushi Made Easy. Sterling Publ. Co., New York, 2001.

Tanaka, S. & S. Sanaka. *The Tea Ceremony*. Kodansha International Ltd., Tokyo, 2000.

Tohyama, H. *Quick & Easy. Sushi Cook Book*. JP Trading, Inc., Brisbane, 1983.

Tokunaga, M. *New Tastes in Green Tea*. Kodansha International Ltd., Tokyo, 2004.

Tsuchiya, Y. & M. Yamamoto. *The Fine Art of Japanese Food Arrangement*. Kodansha International Ltd., Tokyo, 1985.

Wakuda, T. *Tetsuya*. HarperCollinsPublishers, Sydney, 2000.

Yamamoto, K. & R. W. Hicks. *Step-by-step Sushi*. Magna Books, Leicester, 1992.

Yoshino, M. *Sushi. The Delicate Flavor of Japan*. Gakken, Tokyo, 1986.

ON THE SCIENCE OF FOOD AND FOOD PREPARATION

Barham, P. *The Science of Cooking*. Springer-Verlag, Berlin, 2001.

Belitz, H.-D., W. Grosch & P. Schieberle. *Food Chemistry*. 3rd revised edition. Springer-Verlag, Heidelberg, 2004.

Bloomfield, L. A. *How Things Work. The Physics of Everyday Life*. John Wiley, New York, 1997.

Coultate, T. P. *Food. The Chemistry of its Components*. Royal Society of Chemistry, Cambridge, 2002.

Hillman, H. *The New Kitchen Science*. Houghton Mifflin, New York, 2003.

McGee, H. *On Food and Cooking. The Science and Lore of the Kitchen*. Scribner, New York, 2004.

Motokawa, T. Sushi science and hamburger science. *Perspectives in Biology and Medicine* 32, 489-504, 1989.

O'Hare, M. (ed.) *The Last Word*. New Scientist, Oxford Univ. Press, Oxford, 1998.

Snyder, C. H. *The Extraordinary Chemistry of Ordinary Things*. 3. Ed., John Wiley, New York, 1998.

This, H. *Molecular Gastronomy. Exploring the Science of Flavor*. Columbia Univ. Press, New York, 2006.

Wolke, R. L. *What Einstein Told His Cook. Kitchen Science Explained 1 & 2*. W. W. Norton & Co, New York, 2002 & 2005.

ON FATS, NUTRITION, AND WELLNESS

Crawford, D. & D. Marsh. *The Driving Force*. Harper & Row, New York, 1989.

Cunnane, S. C. *Survival of the Fattest*. World Scientific, Singapore, 2005.

Ewin, J. *Fine Wines and Fish Oil: The Life of Hugh Macdonald Sinclair*. Oxford University Press, Oxford, 2002.

Horrobin, D. *The Madness of Adam and Eve*. Bantam Press, London, 2001.

Mouritsen, O. G. *Life – As a Matter of Fat. The Emerging Science of Lipidomics.* Springer-Verlag, Heidelberg, 2005.

Pond, C. *The Fats of Life.* Cambridge University Press, Cambridge, 1998.

Schwarcz, J. *Let Them Eat Flax.* ECW Press, Toronto, 2005.

Taubes, G. The soft science of dietary fats. *Science* 292, 2536-2545, 2001.

Wood, P. A. *How Fat Works.* Harvard University Press, Cambridge, 2006.

ON JAPANESE CULTURE, ESPECIALLY IN RELATION TO FOOD

Bestor, T. C. *Tsukiji: The Fish Market at the Center of the World.* University of California Press, Los Angeles, 2004.

Bramble, P. S. *Culture Shock! Japan. A Guide to Customs and Etiquette.* Times Books Int., Singapore, 2004.

Dower, J. *The Elements of Japanese Design. A Handbook of Family Crests, Heraldy, and Symbolism* (with drawings by Kiyoshi Kawamoto). Weatherhill, New York, 1998.

Ekuan, K. *The Aesthetics of the Japanese Lunchbox.* MIT Press, Cambridge, 1998.

Juniper, A. *Wabi Sabi. The Japanese Art of Impermanence.* Tuttle Publ., Boston, 2003.

Kerr, A. *Lost Japan.* Lonely Planet Publ., London, 1996.

Nihongo, K. & K. Shahen. *Unfolding Japanese Traditions.* Apricot, Tokyo, 1996.

Reichhold, J. *Haiku Techniques.* Frogpond J. Haiku Soc. Amer., Autumn 2000.

Riccardi, V. A. *Untangling My Chopsticks. A Culinary Sojourn in Kyoto.* Broadway Books, New York, 2003.

ILLUSTRATION CREDITS

- Thank you to Sticks 'n' Sushi (Copenhagen) for making available the photographs on pp. 17, 43, 149, 164, 166, 195, 203, 212, 217, 221, 231, and 233. Sticks 'n' Sushi owns the copyright to these images.
- Chef Endo and Chef Watanabe gave permission to include the photographs from Kibune Sushi (Vancouver, Canada) on pp. 6 and 7.
- Takeo in Gothersgade (Copenhagen) has drawn the sushi calligraphy on p. 15.
- Michael Morrissey photographed and made available the pictures of frozen tuna on pp. 77 and 204.
- Thank you to bar'sushi (Odense) for making available the photographs on pp. 22, 162, 170, and 171. bar'sushi owns the copyright to these images.
- The painting on p. 59 was made by Gaute Haugland.
- Musholm Lax made available the photograph of the fish farm in the Great Belt (Storebælt), Denmark on p. 78.
- Ryuusei Matsuo supplied the image on p. 91 from the Drew-Baker Festival.
- The old wooden fermentation room on p. 96 is at the Kikkoman Corporation.
- Malcolm Mackley gave permission to reproduce the micrograph of starch granules on p. 99.
- Jacob Termansen took the pictures on pp. 157 and 160.
- The information campaign 'To gange om ugen' (Two Times a Week) by Food Marketing ApS gave permission to reproduce the photographs of herring, sea bass, pike-perch, and pollock on pp. 210, 211, and 213.
- Paavo Kinnunen took the picture of siika on p. 213.
- Hans Hillevaert gave permission to reproduce the photograph of the Loligo squid on p. 218.
- ND Shii-take gave permission to reproduce the photograph of the shiitake mushrooms on p. 238.
- PrimaFrø gave permission to reproduce the images of purslane on p. 246.
- The patterns which introduce the chapters of the book are taken from classical Japanese stencils which were used for the dyeing of kimonos (BNN, Inc.).
- The classical Japanese woodblock prints which appear at the beginning of each section and several places throughout the book are by Shigemasa Kitao(1739-1820), Utagawa (Ando) Hiroshige (1797-1858), and Katsushika Hokusai (1760-1849).
- All the water colours were painted by Tove Nyberg.

Where no other credits are given, the photographs were taken by the author or by Jonas Drotner Mouritsen, who retain their copyright.

INDEX